Applied Cognitive Ecostylistics

Also available from Bloomsbury

Corpus-Assisted Ecolinguistics, Robert Poole
Storytelling and Ecology, Anthony Nanson
Stylistic Manipulation of the Reader in Contemporary Fiction,
edited by Sandrine Sorlin

Applied Cognitive Ecostylistics

From Ego to Eco

Edited by
Malgorzata Drewniok and Marek Kuźniak

BLOOMSBURY ACADEMIC
LONDON • NEW YORK • OXFORD • NEW DELHI • SYDNEY

BLOOMSBURY ACADEMIC
Bloomsbury Publishing Plc, 50 Bedford Square, London, WC1B 3DP, UK
Bloomsbury Publishing Inc, 1359 Broadway, New York, NY 10018, USA
Bloomsbury Publishing Ireland, 29 Earlsfort Terrace, Dublin 2, D02 AY28, Ireland

BLOOMSBURY, BLOOMSBURY ACADEMIC and the Diana logo are
trademarks of Bloomsbury Publishing Plc

First published in Great Britain 2024
Paperback edition published 2026

Copyright © Malgorzata Drewniok and Marek Kuźniak and Contributors, 2024, 2025

Malgorzata Drewniok and Marek Kuźniak and Contributors have asserted their right under
the Copyright, Designs and Patents Act, 1988, to be identified as Authors of this work.

Cover design: Jade Barnett

All rights reserved. No part of this publication may be: i) reproduced or transmitted in any form, electronic or mechanical, including photocopying, recording or by means of any information storage or retrieval system without prior permission in writing from the publishers; or ii) used or reproduced in any way for the training, development or operation of artificial intelligence (AI) technologies, including generative AI technologies. The rights holders expressly reserve this publication from the text and data mining exception as per Article 4(3) of the Digital Single Market Directive (EU) 2019/790.

Bloomsbury Publishing Plc does not have any control over, or responsibility for, any third-party websites referred to or in this book. All internet addresses given in this book were correct at the time of going to press. The author and publisher regret any inconvenience caused if addresses have changed or sites have ceased to exist, but can accept no responsibility for any such changes.

A catalogue record for this book is available from the British Library.

Library of Congress Cataloging-in-Publication Data
Names: Drewniok, Malgorzata, editor. | Kuźniak, Marek, editor.
Title: Applied cognitive ecostylistics : from ego to eco / edited by
Malgorzata Drewniok and Marek Kuźniak.
Description: London ; New York : Bloomsbury Academic, 2024. |
Includes bibliographical references and index. |
Summary: "This volume is a collection of the latest research that seeks to apply the theory and methodology developed over the last 40 years in the area of applied cognitive ecostylistics to both literary and real-life texts, engaging with a wealth of examples from First World War poetry and Anne of Green Gables through to Condé Nast Traveller hotel descriptions. Exploring the cultural effects of the eco-turn, the collection engages the reader in the problem of the present-day Anthropocene, manifested as Ego-Eco tensions at the level of communicating self-needs and the needs of the Other"– Provided by publisher.
Identifiers: LCCN 2023048289 (print) | LCCN 2023048290 (ebook) | ISBN 9781350362185 (hardback) | ISBN 9781350362222 (paperback) | ISBN 9781350362208 (epub) |
ISBN 9781350362192 (ebook)
Subjects: LCSH: Ecocriticism. | Literary style. | Cognitive grammar. |
Language and languages–Style. | LCGFT: Essays.
Classification: LCC PN98.E36 A67 2024 (print) | LCC PN98.E36 (ebook) |
DDC 801/.95–dc23/eng/20240206
LC record available at https://lccn.loc.gov/2023048289
LC ebook record available at https://lccn.loc.gov/2023048290

ISBN: HB: 978-1-3503-6218-5
PB: 978-1-3503-6222-2
ePDF: 978-1-3503-6219-2
eBook: 978-1-3503-6220-8

Typeset by Integra Software Services Pvt. Ltd.

For product safety related questions contact productsafety@bloomsbury.com.

To find out more about our authors and books visit www.bloomsbury.com
and sign up for our newsletters.

Contents

List of Figures   vii
List of Tables   viii
List of Contributors   ix

Introduction: A New Ecology of Language   *Peter Stockwell*   1

Part One   From EGO: Self-needs, Readership, Society   13

1. From Ego- to Eco-Centricity: Macro- and Micro-levels of Condé Nast Traveller Hotel Descriptions – A Cognitive-Linguistic Account   *Malgorzata Drewniok and Marek Kuźniak*   15
2. A Diffractive Approach to Reader Response (with Reference to Barnes's *The Sense of an Ending*)   *Amélie Doche*   33
3. Cohesion and Solidarity in Covid-related Addresses to the Nation   *Chris Fitzgerald and Helen Kelly-Holmes*   51
4. Using Think-Aloud Data to Explore Pathetic Fallacy's Impact on Narrative Empathy   *Kimberley Pager-McClymont and Fransina Stradling*   69
5. Reader's Reactions to Descriptions of Landscape in Polish Translations of *Anne of Green Gables*   *Beata Piecychna*   99

Part Two   To ECO: Nature, Culture, and Beyond   115

6. Methodological Implications of Building *The Corpus of News on Economic Inequality (1971–2020)*: Text-readable Data vs OCR Material   *Eva M. Gómez-Jiménez*   117
7. Modelling the Landscape of Wilfred Owen's 'Futility'   *Marcello Giovanelli*   135
8. Intralingual Eco-Translation Insights into *Macbeth* in African American Urban Slang   *Michał Garcarz*   153

9   Fictional Ekphrasis Representing Childhood Trauma in M.
    Atwood's *Cat's Eye* *Polina Gavin*   171
10 Body, Mind, and Nature in Rossetti's 'For a Venetian Pastoral by
    Giorgione (In the Louvre)' *Maria-Eirini Panagiotidou*   189

Conclusion: Pathways to Eco *Malgorzata Drewniok* and *Marek Kuźniak*   207

Index   218

Figures

2.1	Propagation of a wave through a chink. 2008. This file is licensed under CC BY 4.0	35
2.2	Diffraction in ocean waves. 2009. © Exploratorium. This file is licensed under CC BY-NC-SA 2.0	36
2.3	Focal resonance between Tony Webster and Reviewer 2 (R2 Chapter 2)	43
4.1	Example 1 of participants' think-aloud responses	78
4.2	Example 2 of participants' think-aloud responses	80
4.3	Example 3 of participants' think-aloud responses	81
4.4	Example 4 of participants' think-aloud responses	82
6.1	*Abbyy FineReader* interface (2022)	126
7.1	The canonical event model	141

Tables

2.1	Intertextuality in R1's review of *The Sense of an Ending*	40
3.1	Dates and word-count of speeches in the Irish Covid Speeches Corpus	53
3.2	Top ten keywords in the Irish Covid Speeches Corpus	56
3.3	Top ten three-word units in the Irish Covid Speeches Corpus	57
3.4	Frequencies of pronouns used in the Irish Covid Speeches Corpus	58
3.5	Concordance of *one of us* in the Irish Covid Speeches Corpus	58
3.6	Occurrences of figurative expressions in the Irish Covid Speeches Corpus	60
4.1	Scoring system used to track participants' perception of PF	77
4.2	Participant's scores of PF perception	83
4.3	Perception of PF and evidence of empathy in think-aloud responses cross-tabulated	87
4.4	List of survey questions	93
4.5	Matrix table for survey question 20	95
4.6	Fernandez-Quintanilla's 'implicit-evidence-for-empathy codes	96
6.1	Overall process for the compilation of the DINEQ corpus	123
6.2	Access to UK national newspapers through *Nexis Uni*	124
8.1	The adaptation technique of equivalence	161
8.2	The adaptation technique of compensation	162
8.3	The adaptation technique of enrichment	163
8.4	The use of gerunds and continuous verbs	165
8.5	The use of conjuncts	165
8.6	The use of contractions	166
8.7	The use of adjectives	167

Contributors

Amélie Doche is an AHRC-funded doctoral researcher in English Language and Literature at Birmingham City University, UK. Her PhD – carried out in collaboration with the literature development agency Writing West Midlands – uses linguistic methods to uncover the ideologies permeating contemporary literary culture. Amélie's research interests include reader-response stylistics, discourse analysis, qualitative methodologies, contemporary book culture and the creative industries. Her most recent published work explores digital literary practices during the Covid-19 pandemic (2022). Prior to her PhD, Amélie gained an MRes (2020) and a BA (2017) in English Studies from Université Jean Moulin Lyon 3.

Malgorzata Drewniok is Head of International College at University of Lincoln, UK. She was awarded a PhD in Linguistics (stylistics) from Lancaster University, UK in 2014, with a thesis on how the change of identity of fictional characters in *Buffy the Vampire Slayer* TV series (1997–2003) is signalled in language. Her research interests include stylistics, pragmatics and language of popular culture. She has published on the language of popular fiction (film, TV and novels). Language of advertising and branding is her more recent interest, and she has been working on a project on the language of luxury hotels.

Chris Fitzgerald's research focuses on exploring oral history documents from a linguistic perspective and aims to promote the potential reciprocity between these two disciplines. He is currently engaged in postdoctoral research at Mary Immaculate College, Limerick, as part of the Interactional Variation Online project, exploring multimodal analyses of virtual meetings.

Michał Garcarz is Professor of Linguistics in the Institute of English Studies at University of Wrocław, Poland, where he teaches and researches various areas of specialized translation and translation environment for specialized purposes, sociolinguistics, lexicography and linguistic data documentation. He is Head of the Department of Translation, a freelance translator, communication for business trainer and language coach, and corpus maker.

Polina Gavin is a doctoral researcher at Aston University, UK. Her research interests centre on a cognitive stylistic approach to cross-media interrelations in ekphrastic writing. She is currently developing a framework that gives insight into ekphrasis as a contemporary creative writing practice and explores its reception by readers.

Marcello Giovanelli is Reader in Literary Linguistics at Aston University, UK. He has research interests in applications of Text World Theory and Cognitive Grammar to literary discourse and in the language of First World War literature. His books include *Text World Theory and Keats' Poetry* (2013), *Teaching Grammar, Structure and Meaning* (2014), *Cognitive Grammar in Stylistics: A Practical Guide* (with Chloe Harrison, 2018), *New Directions in Cognitive Grammar and Style* (with Chloe Harrison and Louise Nuttall, 2021) and *The Language of Siegfried Sassoon* (2022). He has published widely on cognitive stylistics and applied linguistics in major international journals.

Eva M. Gómez-Jiménez is Lecturer in English Language and Literature at the University of Granada, Spain. Her research interests include critical discourse analysis, literary stylistics and grapholinguistics. She is currently working on the diachronic discourse of economic inequality in the UK press. Her recent publications include the book *The Discursive Construction of Economic Inequality: CADS Approaches to the British Media*, co-edited with Prof. Michael Toolan.

Helen Kelly-Holmes is Professor of Applied Languages at the University of Limerick, Ireland. Her research focuses on the interrelationship between media, markets, technologies and languages, with a particular interest in the economic aspects of multilingualism. Recent books include *Language, Global Mobilities, Blue-Collar Workers and Blue-collar Workplaces* (Ed. with K. Gonçalves, 2020); *Sociolinguistics from the Periphery: Small Languages in New Circumstances* (with S. Pietikainen, A. Jaffe & N. Coupland, 2016); *Language and the Media* (2015).

Marek Kuźniak is Professor of Linguistics at University of Wrocław, Poland. He is a certified translator of English. Marek Kuźniak is also Head of the Institute of English Studies and Chair of the Council for Linguistics at the Faculty of Letters, University of Wrocław. He has published in the field of cognitive linguistics, cognitive poetics, legal translation, lexicology, pragmatics and philosophy of language. His latest release is *Geometry of Choice: Language, Culture, and Education* (2021).

Kimberley Pager-McClymont is a researcher and Academic English Skills Coordinator at the University of Aberdeen's International Study Centre UK. Kimberley's PhD is in stylistics from the University of Huddersfield, UK. She focuses on extended emotion metaphors such as pathetic fallacy and their impact on narratives, in literary and multimodal texts. She is Editor for the *Journal of Languages, Texts, and Society* and Associate Editor for the *Cambridge University Press Element Series* in cognitive linguistics. She is also Webmaster for the Poetics and Linguistics Association.

Maria-Eirini Panagiotidou is Associate Professor of Linguistics at West Chester University of Pennsylvania, USA. She holds a PhD in linguistics from the University of Nottingham, UK, and has taught at Virginia Commonwealth University and the University of Maryland, University College. Her research interests and publications lie in the areas of literary linguistics, cognitive poetics, iconicity, ekphrasis and intertextuality. Her recent publications include *The Poetics of Ekphrasis* (2022).

Beata Piecychna is Assistant Professor at the Faculty of Philology, the University of Białystok, Poland. She has published papers on translational hermeneutics, philosophy of language, translation pedagogy and cognitive translatology. Her research interests include translational stylistics, philosophical hermeneutics, the hermeneutics of translation, translation theory and embodied aesthetics. She was a visiting scholar at the University of Cambridge (2019) and at the University of California in San Diego (2020).

Peter Stockwell is Professor of Literary Linguistics at University of Nottingham, UK. He specializes in literary linguistics, cognitive poetics, stylistics, applied linguistics, science fiction and surrealism. He is also interested in sociolinguistics and language education. He has published extensively in international journals and edited volumes. His recent book publications include *Cognitive Poetics: An Introduction* (2nd ed. 2020) and *The Language of Surrealism* (2017).

Fransina Stradling is a PhD student in cognitive stylistics at the University of Huddersfield, UK. Her research explores the role of language in eliciting reader empathy with fictional characters, particularly in Modernist short stories. Fransina currently co-hosts bi-yearly interdisciplinary discussion events about research into empathy and reading. In 2020 she won the Palgrave Prize for best written version of a student paper at the PALA 2019 conference with her paper on empathetic engagement with the mind of the protagonist in 'The Story of an Hour'.

Introduction: A New Ecology of Language

Peter Stockwell

Self-centred stylistics

Traditionally, the discipline of stylistics has been regarded as being tightly focused on textual form, while excluding matters of authorial intention and creativity on the one hand and readerly effect and interpretation on the other. The social, cultural, biographical and historical context was left to the literary critics and social theorists. Although this characterization was never entirely true, the focus on the text itself tightened the discipline around an analytical practice that was literally self-centred: the analysing reader and the text in hand. The stylistician presents themself as a common reader, a normal reader, one whose responses and sensitivities are not esoteric nor abstruse but make a claim to demotic ordinariness. The stylistic reader is the intelligent reader, with an implicit and complicit appeal to the shared intuitions of the audience.

In this sense, traditional stylistic practice can be seen as egocentric, but not explicitly so. The manifestation of the averageness and ordinariness of the projected reader serves to erase the presence of a reader altogether as if there is only a text and a set of disembodied readings that arise from it. This sounds as if I am disparaging, and I am not intending to be. This phase of stylistics produced a great understanding of the linguistic techniques and patterns by which a text could be powerful, persuasive, manipulative, striking or inspirational. For the most part, stylistic accounts were indeed recognizable and plausible to a wider readership, and stylistics itself as a discipline managed to retain this connection with the wider population, just as other branches of critical theory and scholarly practice were becoming more and more divergent from the reading practices and interests outside the academy.

In any case, in practice, these constraints have always been more permeable than they might have appeared. A linguistic description of a text is rarely simply and purely descriptive, without motivation, commentary or a discussion of its significance. And as the methodological toolkit of stylistics has expanded over the last half century or more, so the capacity for rigorous stylistic analysis of all of these 'contexts' has also expanded (see Carter and Stockwell 2008 for a representative history).

Developments in cognitive linguistics in particular, in the last two decades, have provided stylisticians with the means to address matters of both productive intention and the situated effects of real texts in the world. Applied cognitive stylistics brings a systematic set of principles to the exploration of the entire rhetorical environment of real-world discourse. Many of the key insights in the discipline were developed in relation to the reading of literary texts, so the affective impact of such discourse has long been a key concern. Cognitive stylistics has also, however, been interested in authorial and creative stylistic choices, with a critical theoretical eye on the limits of ascribing intentionality to literary authors. Both of these dimensions offer a greater degree of subtlety to an analysis of texts beyond the literary than can be achieved by traditional critical discourse analysis.

The chapters of this book exemplify this rich environment. Any text which lies at the centre of the descriptive account is nevertheless conditioned by the psychological and socio-cultural fields of influence that sustain its core discourse. Applied ecostylistics – such as those exemplified in the following pages – offers a rigorous, integrated and productive means of exploring and explaining how these texts work and how their discursive practices achieve their effects.

From the widest perspective, we can see two ways in which an *ecostylistics* could be manifest. The first practice might be considered a fairly conventional application: the use of stylistic methods to illuminate texts that have ecological concerns at their heart. And, of course, contrastively, texts which deny, avoid or are actively antagonistic to issues of ecology or sustainability can be revealed by a stylistic analysis. We can imagine a stylistic analysis of a nature or landscape lyric, a climate dystopia or an exploration of a text that conspicuously denies ecological concerns as being examples of each of these. In this sort of approach, though, there is nothing inherently ecological about the linguistic models being used. We might see this conventional practice as simply traditional stylistics in the service of a green agenda. In a similar way, for example, a 'feminist stylistics' has nothing inherently feminist in its linguistic model – often, for instance, the transitivity relations within a systemic-functional approach are analysed conventionally to reveal the power, agency and ideological positioning in a text, and thus to argue persuasively for a feminist political framing, but there is nothing essentially feminist in the actual methodology here.

In the same way, an ecological stylistics might draw on conceptual metaphor theory to illuminate the 'stories we live by' (Stibbe 2021). This is still a conventional application: in this case, the cognitive-linguistic model of framing and conceptual metaphor as a means of understanding the extended, structuring metaphorical mappings of public discourse on the environment. Most of the insightful analyses in Virdis' (2022) *Ecological Stylistics* are explorations of key or significant texts about nature and the environment, but the analytical tools (exploring metaphor, foregrounding, point of view, lexical semantics and so on) are conventional. For me, this 'ecological stylistics' is again stylistics in the service of environmentalism rather than being a new type of stylistics itself.

These examples of conventional stylistic analysis in the service of a green ideology are not trivial; on the contrary, they are essential and profound. However, a second, more radical sense of ecostylistics would involve an ecological alteration of the methodology itself. This is not merely an application of stylistics but a re-imagining and recasting of it along different, integrative principles. This would be an ecostylistics that represented the next step in the evolution of stylistics itself, rippling outwards from an analytical practice centred on self-consciousness.

Ripples outwards

How, then, might a genuine methodological shift in stylistic practice be effected? For a comparable moment, we need to look back at some key shifts in literary stylistics in the past. For example, as Shattock (2013) points out, the emergence of the 'literary critic' in the nineteenth century was marked principally by the abandonment of anonymity in book-reviewing, in the form of signed articles in journals and periodicals appearing from around 1860 onwards. This shifted the analysis of literature from a decentred and apparently objective position to an egocentric basis and established the persona of the literary critic in the modern period. Much of the critical account in these times revolved around the moral purpose of literary art (see Habib 2013), with any attention paid to matters of language being drawn from the classical rhetorical tradition.

At the beginning of the twentieth century, however, we can identify a more linguistically focused practice in the hands of those who drew on the sciences of philology and modern rhetoric. The 'formalists' based in Moscow,

St Petersburg and Prague described what they did as 'literary linguistics', and their primary concern was not so much literary criticism as literary 'poetics', and the identification and description of literariness (see Sotirova 2016). Aspects of a new science of language drew on emerging disciplines and approaches such as functionalism, psychology and sociology, and branches of linguistics – such as phonology – that had become possible because of new recording technology. As a result, the discussions by scholars in this tradition looked radically different from the literary critical practices that had gone before.

We might identify another disjunctive shift in methodology with the rise of modern linguistics between Chomsky (1957) and Halliday (1985). The former established a means of formalizing linguistic description, and the latter had a more direct influence on the lexico-grammatical tools that were available for textual analysis. The period between the two can be identified as the emergence of modern stylistics. Other developments that can be characterized as moments of disjunctive progress would, I think, include developments in pragmatics in the 1970s, or sociolinguistic innovations into the 1980s, or the advances in corpus linguistic methods in the 1990s. Each of these changed stylistics itself in radical ways. Crucially, one of the key indicators of the shift can be seen in the different sorts of literary works that came within range of these new stylistics. Systemic-functional linguistics encouraged stylisticians to regard clausal patterns as being meaningful in their own right. Pragmatic advances such as speech-act theory, politeness and viewpoint allowed a new analysis of text as discourse, character interaction and interpersonal relationships in fiction. Conversation analysis and methods from sociolinguistics allowed stylisticians to move on from lyric poetry to prose fiction and into drama and performance. Aside perhaps from 'literary semantics', 'literary pragmatics' or 'pragma-stylistics' (Eaton 1978; Sell 1990 and Hickey 1993, respectively), these radical reformulations at the methodological level were not distinguished by new disciplinary titles. The broad church of stylistics simply assimilated them as part of its radiating circumference.

An exception might be the field of 'corpus stylistics', which I have always thought is actually a misnomer. The key innovation in the use of large bodies of digitized literary texts as part of a tool for stylistic analysis is not so much the corpora themselves as the computer software that allows them to be interrogated and analysed: concordance searches and filters, collocation analyses, the exploration of keywords and clusters. These technological innovations in methodology represent a radical extension in stylistic practice. Essentially, they also represent a shift from egocentric analysis towards an intersubjective palette of comparison. A general reference corpus of English usage in the speech community offers a sort of idealization by massive generalization of 'the reader'. Corpus stylistics – among other things – allows you to check against a backdrop of thousands of users whether your own intuitions about language patterns are indeed normal (see McIntyre and Walker 2019).

My own research interests over the past three decades align more closely with what I think is the dynamic of the chapters in the present collection. The 'cognitive turn' in arts and humanities research represents another disjunction in linguistics that has had a profound effect on stylistics. Both 'cognitive poetics' (echoing the Russian and Czech formalists) and 'cognitive stylistics' are terms that denote a new methodology for the stylistic analysis of literary works. Drawing on a new paradigm of cognitive linguistics and cognitive psychology, cognitive poetics offers another way of generalizing away from an egocentric focus: the foundational models and principles make a claim to the capacities of mind of readers in general. Where corpus stylistics tests egocentric intuitions out with reference to multiple readers, cognitive poetics treats the self as a case study within humanity. The claim is that your experiences and responses as a reader are valid because you share cognitive capacities with all other readers (see Stockwell 2020).

A criticism levelled at early cognitive poetics was that it privileged psychology over sociolinguistics, preferring a sense of mind over a material sense of the social and cultural situation. I think this was a mistaken view, but most recently, the field has developed the notion of '4E cognition' that offers a particularly pertinent set of concepts for an ecological stylistics that is more radical than simply an application of stylistics to green-concerned texts. Cognitive poetics was founded on the principle that cognition is *embodied*, in the sense that the mind is not circumscribed by the brain but also by all the body's perceptions and our physical experience of the world. We perceive and experience the world and our lives largely on the terms of our bodily condition as humans. Our cognitive sensorium is not limited to our own body either, since much of our language and thought is dependent on and reactive to and predictive of other people in our vicinity. In the sense that our mind also consists of our modelling of other minds, cognition is *extended*. We are implicated in each other. This view of cognition is also *enactive* in that it is dynamic and arises from the interaction with our environment. And lastly, cognition is *embedded* in our social and cultural environment (see Tewes, Durt and Fuchs 2017, and Newen, De Bruin and Gallagher 2018).

These four Es represent a set of experiential parameters that present language use as an inter-subjective mind being situated. The 'situatedness' (Barsalou

2009; 2016: 15–16) is imagined from the local proximity of other minds in the immediate vicinity, right out to the general sense of other people in society and culture. To the extent that these more distant minds tend to be grouped and characterized together, Palmer (2004) has called them 'social minds'. To capture the sense that all of these different aspects of cognition are integrated and interanimate each other, we might informally add a fifth 'E': *ecology*. Indeed, the ecology of language might be regarded as a super-dimension, in the sense that it encompasses the notion of our cognitive situation as embodied, extended, embedded and enactive, and with all of these threads entwined together (see also Rietveld et al 2018). A stylistics that is methodologically driven by 4E cognition would be well on the path towards becoming a radical ecological stylistics.

These arguments can be made more demonstrably, true to the stylistic tradition, by turning to a brief analysis of a literary text. The following simple poem, originally untitled, is from Samuel Taylor Coleridge's Notebook 24, p.78, probably written in May 1814 (see Coburn [1973: 4194] for the annotated manuscript notes and Richards [1977: 158] for the final text).

> Sea-ward, white gleaming thro' the busy scud
> With arching Wings, the sea-mew o'er my head
> Posts on, as bent on speed, now passaging
> Edges the stiffer Breeze, now, yielding, drifts,
> Now floats upon the air, and sends from far
> A wildly-wailing Note.
>
> Samuel Taylor Coleridge, composed 1814

It is a nature poem. As is often typical in Romantic verse, the observing poet appears within the text, and here is drawn to perceive significance in what is depicted: in this case, a seagull ('sea-mew') flying overhead. A traditional stylistic analysis of this text might perhaps focus on the ways the enjambed lines roll on, iconically representing the fleeting bird passing over at speed. All six lines are a single trailing sentence, with a long locative and descriptive adverbial preceding the 'sea-mew' itself as the subject, with its main verb 'Posts on' beginning the sequence of present tense and present participle verb forms across the following lines. The momentum effect of the enjambment is cleverly set up contrastively without breaking the syntactic enjambment of the first two lines by having both those lines ending in the terminal voiced /d/ of 'scud' and 'head'. The line endings after that, as the seagull speeds away, are more open, less voiced and vocalic. There is also strong alliteration throughout (which reminds me of Old English verse like *The Seafarer*, given this context). I particularly notice the alliterative

/w/ in the first two lines, which fades away as the seagull flies and then is strongly sent back with the cry in the final half-line.

A cognitive poetic analysis of this might draw on Cognitive Grammar (Langacker 2008; Harrison et al. 2014; Giovanelli and Harrison 2018) to capture the experience of perspectivization. This would focus on the movement across the linked clauses, with the momentum of the trajectory accelerating, then slowing, falling, floating and ending at a still point. The observer's vantage point describes a sequential scan – in which the linguistic articulation denotes step-by-step the arc of the bird's movement. (A summary scan would be something like: 'The bird flew away'.) The vantage point is initially static, but (with 'Posts on') moves suddenly very fast ('bent on speed'), before then appearing gradually ('passaging/ Edges') to slow down ('drifts … /floats') and become still at a distant point ('from far'). This sequential trajectory iconically captures the observer's experience on the ground as the bird flashes overhead before the distant perspective seems to slow down in relative terms. The sequentially scanned motion of the bird is rendered as an objective construal: the observing consciousness is fixed on the bird flying away, with only 'o'er my head' to remind us of the poetic persona.

Cognitive Grammar is a methodology that aligns with 4E cognition; it could arguably be regarded as being relatively more ecological as a framework than, say a traditional syntactic analysis, as a result of this. Perhaps an even more radically ecological stylistics could be achieved by a fully 4E framing that involved tracing the mind-modelling around this short literary work (Zunshine 2006; Stockwell 2022). For example, there are three embedded minds that need to be monitored during a reading of this poem: the authorial Coleridge, the poetic persona who appears as the bird-watcher within the poem and the seagull itself.

The poetic persona's mind is occupied almost completely by the flight of the bird. Their perception is wholly visual and vivid (perhaps even cued up subliminally by the sound of 'Sea-ward' as 'see-word' if read aloud), so that the final switch to the sound of the gull's cry is contrastively striking. The seabird is single-minded, keeping to its swift trajectory overhead and away; it is a natural part of the land and seascape, with only a hint ('bent on speed') that it has intentions and consciousness. The poem cleverly blends both the perspective of the person on the ground and the flying bird. The sequence of verbs and progressive verb-forms is described by the watcher; but the present tense nature of these, and the breathless fragmented syntax seem more a mirroring of the bird's cognition rather than the human's. The authorial Coleridge is the organizing mind here. The watcher on the ground hears the seagull's 'Note', but the Romantic poet receives that 'Note' (also capitalized and placed for significance) as a message and perhaps there is a phonic echo of 'Samuel' in 'sea-mew'. The poem ends, though, with a half-line, and feels to me ambivalently finished (in the definite terminal plosive /t/) and yet unfinished. In the gesture towards a message that is unstated, of course I am reminded of the albatross, another seabird and its ambiguous meaning in Coleridge's most famous poem, *The Rime of the Ancient Mariner*.

I would argue that taking a mind-modelling approach from 4E cognition is methodologically more ecological as a stylistic approach because the linguistic framework itself is integrative and interconnected. 4E cognition is methodologically ecological. What, though, would be the possibilities of moving beyond even this to find the limits of what is possible?

The limits of ecostylistics

A thoroughly ecological linguistics, integrated, networked and reaching into both individual psychology and into social and cultural space – and then extended diachronically across history – is not yet available. Perhaps one step towards this goal is a radical re-imagining of discourse such as that suggested by Doreen Massey in her philosophy of space. Massey (1994; 2005) replaces the discrete and static social variables of age, sex, gender, wealth, class, political outlook, race, ethnicity and so on not simply with a vague intersectional sense that would nevertheless preserve these categories, but with an innovative perspective on the mind and society based on space. For Massey (2005: 9), space is the product of interrelations, is concerned with multiplicity and heterogeneity, and is always dynamically under construction. Massey's concern in articulating society and culture through the primacy of space (rather than through time by privileging history as a single thread – though it encompasses it) is political. Furthermore, you alter space by being in it. An analytical description of the place you are in, and the sequence of moving through it, and the relative positions of other people in their own spaces, is a description that is fundamentally and radically ecological. If it was ever to be fully developed, it would genuinely be a new ecology of language (originally envisaged by Haugen 1972).

For example, the Coleridge poem above uses positions in space to capture speed and time, and transcend distance. The sequence of moments is articulated in present tense verbs 'Posts ... Edges ... drifts ... floats ... sends', interspersed with continuous forms 'gleaming ... arching ... passaging ... yielding'. These all present motion by describing positions in the air, and though the sequence across the lines will always be read as a progression across time, the fragmented syntax is also interspersed not with 'then', or 'next', but with a reiterated 'now … now … Now'. We are being placed into the seabird's experience through an expression of immediacy. This takes us away with the bird until we are jarringly repositioned by 'sends from far', which deictically puts us back on the ground with the poetic persona.

The position of the static observer in the poem lies between the accelerating movement of the seabird and the echoic sympathetic repositioning of the immersed reader, almost literally carried away until we are brought back first to the poetic persona's position and then out of the poem altogether in the last line. The unsaid second half of that line, and the unstated meaning of the seabird's message, relocates us back in the world we share with Coleridge. Like him (or rather, our mind-modelled sense of him), we understand that there is a profound message in the seabird's flight and cry, but we are unable to comprehend it fully. The best we can hope for is to have shared the experience which the poem enacts in us.

There are risks in testing the limits of a fully fledged ecostylistics along these lines, of course. One is that such an 'ecosophy' is prescriptive of an ideological position – and prescriptivism is a framing that stylistics has fundamentally resisted throughout its history (see Virdis' [2022: 16] comments on Naess' [1995] use of the term 'ecosophy'). Second might be that an integrated and fully contextualized stylistics is so complicated that it defies simple accessibility – and plain-speaking and the avoidance of obscurity has been a long-standing badge of honour of the discipline. Third, an historical and progressive shift to communalism in response risks the effacement of individual perception and perhaps idiosyncratic interpretation that can often be joyously persuasive when framed with text-stylistic evidence. And lastly, a thorough ecological stylistics seems to require a paradigm shift in linguistics, certainly as it is institutionally situated and practised – and this is a very large demand indeed.

I started this introduction by noting the different manifestations of ecologically oriented stylistics, beginning with the historical development of the discipline that widened the field out as a richer set of linguistic tools became available. Ecostylistics can be conventional (applying stylistic methods in the service of environmentalism), or it can be more radical (adopting methods that in themselves embody an ecology of language). The present challenge for ecologists of language lies in specifying exactly what the stylistics of a thoroughly radical ecological analysis would contain, but as stylisticians, we are already particularly good at this, a fact to which the chapters in this book can attest.

References

Barsalou, L. W. (2009), 'Simulation, situated conceptualization, and prediction', *Philosophical Transactions of the Royal Society B* 364: 1281–9.

Barsalou, L. W. (2016), 'Situated conceptualization: Theory and applications', in Y. Coello and M. H. Fischer (eds), *Foundations of Embodied Cognition, Vol. 1: Perceptual and Emotional Embodiment*, Hove: Psychology Press, 11–37.

Carter, R. and P. Stockwelleds (2008), *The Language and Literature Reader*, London: Routledge.

Chomsky, N. (1957), *Syntactic Structures*, The Hague: Mouton.

Coburn, K., ed. (1973), *The Notebooks of Samuel Taylor Coleridge. Vol 3: 1808–19*, London: Routledge & Kegan Paul.

Eaton, T., ed. (1978), *Essays in Literary Semantics*, Heidelberg: Julius Groos.

Giovanelli, M. and C. Harrison (2018), *Cognitive Grammar in Stylistics*, London: Bloomsbury.

Habib, M. A. R., ed. (2013), *The Cambridge History of Literary Criticism (Vol. 6: The Nineteenth Century)*, Cambridge: Cambridge University Press.

Halliday, M. A. K. (1985), *An Introduction to Functional Grammar*, London: Edward Arnold.

Harrison, C., L. Nuttall, P. Stockwell and W. Yuan, eds (2014), *Cognitive Grammar in Literature*, Amsterdam: John Benjamins.

Haugen, E. (1972), *The Ecology of Language: Essays by Einar Haugen*, ed. A. S. Dil, Stanford: Stanford University Press.

Hickey, L. (1993), 'Stylistics, pragmatics and pragmastylistics', *Revue Belge de Philologie et d'Histoire* 71 (3): 573–86.

Langacker, R. (2008), *Cognitive Grammar: A Basic Introduction*, New York: Oxford University Press.

McIntyre, D. and B. Walker (2019), *Corpus Stylistics: Theory and Practice*, Edinburgh: Edinburgh University Press.

Massey, D. (1994), *Space, Place and Gender*, Cambridge: Polity Press.

Massey, D. (2005), *For Space*, London: SAGE.

Naess, A. (1995), 'The shallow and the deep, long range ecology movement: A summary', in A. Drengson and Y. Inoue (eds), *The Deep Ecology Movement: An Introductory Anthology*, Berkeley: North Atlantic Books, 3–9.

Newen, A., L. De Bruin and S. Gallagher, eds (2018), *The Oxford Handbook of 4E Cognition*, Oxford: Oxford University Press.

Palmer, A. (2004), *Fictional Minds*, Lincoln: University of Nebraska Press.

Richards, I. A., ed. (1977), *The Portable Coleridge*, Harmondsworth: Penguin.

Rietveld, E., D. Denys and M. van Westen (2018), 'Ecological-enactive cognition as engaging with a field of relevant affordances: the skilled intentionality framework (SIF)', in A. Newen, L. De Bruin and S. Gallagher (eds) The Oxford Handbook of 4E Cognition, Oxford: Oxford University Press, 41–70.

Sell, R., ed. (1990), *Literary Pragmatics*, London: Routledge.

Shattock, J. (2013), 'Contexts and conditions of criticism 1830–1914', in M. A. R. Habib (ed.), *The Cambridge History of Literary Criticism* (Vol. 6), 21–45. Cambridge: Cambridge University Press.

Sotirova, V. (2016), 'Introduction', in V. Sotirova (ed.), *The Bloomsbury Companion to Stylistics*, 3–18. London: Bloomsbury.

Stibbe, A. (2021), *Ecolinguistics: Language, Ecology and the Stories We Live By* (2nd edn). New York: Routledge.

Stockwell, P. (2020), *Cognitive Poetics: An Introduction* (2nd edn). London: Routledge.

Stockwell, P. (2022), 'Mind-modelling literary personas', *Journal of Literary Semantics* 51 (2): 131–46.

Tewes, C., C. Durt and T. Fuchs, eds (2017), *Embodiment, Enaction, and Culture: Investigating the Constitution of the Shared World*, Cambridge: MIT Press.

Virdis, D. (2022), *Ecological Stylistics: Ecostylistic Approaches to Discourses of Nature, the Environment and Sustainability*, Basingstoke: Palgrave Macmillan.

Zunshine, L. (2006), *Why We Read Fiction: Theory of Mind and the Novel*, Columbus: Ohio State University Press.

Part One

From EGO: Self-needs, Readership, Society

The focus of the first part of the book is on the EGO and the start of the journey towards ECO. Chapter 1 (Drewniok and Kuźniak) and Chapter 3 (Fitzgerald and Kelly-Holmes) both look at the discoursal shift linked to the recent pandemic. Chapter 1 examines short texts (blurbs) in the form of hotel descriptions in *Condé Nast Traveller* magazine, tracing the change of focus from EGO to more ECO. When the two small corpora were compared, the microlevel style (lexical and grammar choices) remained unchanged, but the macrolevel focus shifted from traveller-centric to traveller being part of a wider ecosystem. The analysis is kept in a cognitive-linguistic spirit with a particular emphasis on the insights from Kövecses's (2020) Extended Conceptual Metaphor Theory. Chapter 3 has a more serious focus: the Covid-related discourse in the communication to the nation coming from the Irish government. Christopher Fitzgerald and Helen Kelly-Holmes discuss how figurative language has been used to appeal to solidarity among the Irish society.

In contrast, the other three chapters in this section of the book deal with readers – both in terms of analytical tools (Chapter 2) and reader responses to texts (Chapters 4 and 5). In Chapter 2, Amélie Doche proposes a novel, diffractive approach to analysing readers' responses on the basis of online reviews of *The Sense of an Ending* by Julian Barnes (2011). Doche shows that diffraction can account for conflicting variables such as readers' reviewing practices and the researcher's relationship with data.

Kimberley Pager-McClymont and Fransina Stradling discuss pathetic fallacy's impact on narrative empathy of the reader in Chapter 4. They argue the benefit of using reader data to examine the experiential impact of conceptual metaphor mappings on empathy and emphasize the importance of using the reader data in stylistic analyses in general. Pager-McClymont and Stradling show how conceptual metaphor can explain how a textual trigger helps readers create a mental representation.

Similarly, in Chapter 5, Beata Piecychna examines implied reader's reactions to landscape descriptions in the Polish version of *Anne of Green Gables* (L. M. Montgomery 1908). She explores how selected fragments of the translated work create a specific aesthetic experience for the reader. Piecychna's chapter is also an excellent showcase for the translational stylistics approach.

1

From Ego- to Eco-Centricity: Macro- and Micro-levels of Condé Nast Traveller Hotel Descriptions – A Cognitive-Linguistic Account

Malgorzata Drewniok and Marek Kuźniak

Introduction

Travel magazines are written in a very specific style and incorporate various types of texts. In this chapter, we will look specifically at hotel descriptions via a cognitive-linguistic lens. The chapter has two parts to it. In the first part, we would like to look at the problem from a more theoretical angle. This means excursions into selected philosophical and socio-cultural issues that probe the ontological-epistemic continuum with EGO on the one end of the spectrum of the life experience and ECO on the other. We will particularly focus the reader's attention on the EGO and how it practically extends beyond itself to reach the ECO aspects of life. In this respect, our paper is preserved within the anthropocentric research tradition, as the point of departure is a human being.

However, in this discussion, we will strive to capture what appears plainly elusive at first sight, namely, the shift from the strongly EGO-centric perspective into what we call a 'weak' EGO-centric viewpoint, or, better, a non-anthropocentric position, as part of our key argument. we think that this non-anthropocentric attitude lies at the heart of the contemporary discourse about the human individual and their encounters with the aspects of the other, here, nature. Of course, the concept of the other is extremely rich in its signification; it involves all aspects of reality other than the human conceptualizer, whether organic or inorganic, natural or artefactual. We thus necessarily select only one facet of that multi-layered substance, that is, the human domain of resting and how this domain is rendered in the present-day popular travel-oriented magazines in view of the EGO–ECO continuum signalled above.

In the second part of the chapter, we will then try to approach the problem of the observed shift from EGO to ECO-centricity, by examining the authentic excerpts from Condé Nast Traveller. We explore these short texts in more detail, diachronically and synchronically. Since the outbreak of the pandemic, the travel industry has been forced to change. Has *Condé Nast Traveller* changed too? We will compare a small corpus of hotel blurbs from the magazine from 2019 and from 2021 to see whether the choice of the hospitality venues and how they are described have changed. We will focus on lexical and grammar choices to define the specific style of such blurbs and any changes between pre-pandemic and (post)pandemic language.

PART ONE

Cognitive-linguistic approach as a pathway to ECO

We explore the issues in question through the lenses of a cognitive-linguistic approach. In its experientialist philosophy (see Lakoff and Johnson 1999; Evans and Green 2006), cognitive linguistics (CL) seeks to bridge the gap, or in other words, overcome the subject conceptualizer and the world out there as an object of conceptualization. In this regard, the CL project, in its promotion of individual ontology (see Kardela 2006), can be seen as analogous to eco-philosophical realism as understood by Arne Naess (1989), especially the idealist position. Magdalena Hoły-Łuczaj (2018: 64), in her comprehensive socio-philosophical account of the roots of non-anthropocentrism, elaborates on the idea of Naess's views, which are summed up as three fundamental positions in respect of the natural world:

1. Homocentrist. The power of human imagination is overwhelming. There is no limit to what human genius can project into nature. This is evidenced by the richness of tree symbolism. A few facts are enough for the imagination: the leaves are green, and the shoots grow upward … The rest is a fantastic projection of the human mind.
2. Idealist. Strictly speaking, leaves are not green. Their atoms are colourless; they are not even grey. Therefore, there is a realm beyond the material world. The human mind is in direct contact with this realm and brings this order to nature, elevating it to this actual realm.
3. Eco-philosopher. The richness and exuberance of reality! How great! The appearance of a tree is the basis for a probably infinite number of concretizations in experience![1]

We argue that cognitive linguistics occupies the area between idealism and eco-philosophy. This position marks a particularly subtle shift of perspective in language studies: on the one hand, CL claims universalist ideas in that certain fundamental concepts like, for example, those connected with spacetime designation, such as UP-DOWN, LEFT-RIGHT, FRONT-BACK, CENTRE-PERIPHERY, LINEARITY or BALANCE, all seem to derive from basic human experience that stems from a specific bundle of cognitive-perceptual predispositions, which have been shaped through biological evolution (cf. Johnson 1987; Lakoff 1987; Krzeszowski 1997; Kövecses 2020; Kuźniak 2021). Though rooted in the tradition of the so-called perspectival anthropocentrism,[2] CL reaches out for the ecological (environmental) properties as these have a powerful impact on how a human conceptualizer perceives the world and interacts with it.

Languages studies and deep ecology

Interestingly, the observed shift of language studies from an EGO-centric to ECO-centric perspective is also noted in the latest research, for example, the 2022 edited collection by Marta Bogusławska, Alina Andreea Dragoescu Urlica, Lulzime Kamberi under the much-telling title *From Cognitivism to Ecologism in Language Studies*, though the ambitions outlined in the book definitely exceed beyond the standardly conceived linguistics. The goals formulated therein by, for example, Bogusławska (2022) may be compared to the eco-philosophical position known as *deep ecology* in contrast with *shallow ecology*.[3] The term 'deep ecology' first appeared in 1973, in Naess's article 'The Shallow and the Deep, Long-Range Ecology Movement'. Here Naess outlined the project of 'deep' ecology, contrasting it with the dominant, 'shallow' ecology movement. As such, shallow ecology supports solutions, such as recycling, that would address the deepening ecological crisis. In doing so, however, shallow ecology, as Naess points out, avoids questions about the source of the ecological crisis. It deals only with ad-hoc attempts to address its symptoms. In Naess's view, on the other hand, a way out of the ecological crisis will only be possible when the world view that caused it is challenged. This is precisely the task that deep ecology sets itself. The change that deep ecology seeks, therefore, is not so much a reform of certain human behaviours as a radical transformation of the attitude that humans take towards the world around them (Hoły-Łuczaj 2018: 25–6). As Hoły-Łuczaj (2018: 9) claims, deep ecology represents a radical strand of eco-philosophy that emerged in the 1970s. Its goal is to remodel ideas about man's place in the world, which will result in the adoption of pro-ecological patterns in human action. In this respect, the position assumed in our chapter owes more to the tradition that may be situated in between the shallow and deep ecology approach.

Deep ecology, cognitive linguistics and the Great Chain of Being

A good illustration of this linking is again CL with its adoption of the traditional Great Chain of Being as a source of ethical and aesthetic human behaviour in relation to other entities. A comprehensive philosophical-historical inquiry into the concept of the Great Chain of Being is provided by Arthur O. Lovejoy (1971). The best-known implementation of the idea of the Great Chain of Being in CL studies is offered by Tomasz P. Krzeszowski (1997: ch. III). Krzeszowski (1997: 64–6) adheres to the fundamental hierarchical ontology of beings by claiming that the Chain represents the core of the Western philosophy of thought as best enshrined by the Judeo-Christian civilizational heritage. According to this ontology, the highest level is occupied by God, then right below, there are humans, then animals, plants and finally, at the very bottom of the hierarchy, inorganic entities are situated.

By applying a cognitive-linguistic apparatus and its philosophy of experientialism, Krzeszowski smoothly proceeds from endo-linguistic phenomena to exo-linguistic facts, such as discourse phenomena (1997: ch. XI). The key vehicle for this spectacular transfer from ego to eco perspective is a conceptual metaphor (ch. VIII), strictly speaking, its elasticity to move up and down the ontological hierarchy of beings, which is reflected in language. For example, humans may metaphorically refer to other humans by reaching out for the highest divine level when they complement one another. They may, however, use the levels below the human to express more negative meanings; for example, a speaker may describe the other person as an animal or a thing. In such cases, the valuation conveyed through the metaphor tends to be negative. And conversely, entities occupying the lowest levels of the hierarchy may be elevated; for example, inorganic entities, plants or animals may be metaphorically deified or anthropomorphized. In other words, a conceptual metaphor turns out to be a powerful tool by means of which we may achieve communicative eco-spherical egalitarianism.[4]

Another thing that links CL and deep ecological movement lies in its focus on various clashes stemming from the life hierarchy implied in the ontological hierarchy of beings. Krzeszowski argues that these conflicts are axiological in nature (they concern the valuative aspects of beings). Deep ecologists claim, in addition, that the hierarchy of the value of life (its degree) is not the only possible criterion in resolving the conflict of interests of different entities. Rather, Naess considered that it should be the aforementioned vital needs[5] (vital needs) and proximity (nearness). Of these two rules, proximity takes precedence because of the special relationship that occurs between people as humans (Hoły-Łuczaj 2018: 48). It is, thus, the notion of CLOSENESS implied in the NEAR–FAR schema and the concept of need implied in the possession schema that appear to set out a particularly sensitive scene, where the EGO-ECO tensions are actuated. As we have already stressed above, we see the conceptual metaphor operating in the context of concrete, authentic textual manifestations best embodied by the Extended Conceptual Metaphor Theory (Kövecses 2020) as an adequate vehicle for capturing these subtle tensions.

Metaphor in context

Zoltán Kövecses's study on metaphors has recently been crowned with his Extended Conceptual Metaphor Theory (ECMT), which sums up many years of highly impactful research in the field (see Kövecses 2006; 2015; 2017; 2020). In his recent 2020 release, the ECMT investigative framework becomes refined and elaborated. One of the major strands along which ECMT is shaped relates to the working of a conceptual metaphor in context. A fundamental question is: why is a particular set of metaphors used in a particular context? In trying to address this issue, Kövecses (2020: ch. IV) distinguishes a schematicity hierarchy, which differentiates levels of metaphor conceptualization depending on the degree the said schematicity, that is, the degree of generalness at which a particular metaphorical construal can be identified. Kövecses argues for four such levels: (a) image schema level, (b) domain level, (c) frame level and (d) mental spaces level.

As Kövecses (2020: 151) puts it, 'a schematicity hierarchy is established that serves as a conceptual pathway from the meaning expressed at the mental-space level all the way to the image-schema level'. Kövecses (2020) further illustrates that conceptual pathway from general to specific as follows:

1. Image schema level: ACTIVITY IS MOTION
2. Domain level: COMMUNICATION IS SELF-PROPELLED (FORWARD) MOTION
3. Frame level: CONFESSIONS ARE RACES

4. Mental-space level: CONFESSIONS ARE BICYCLE RACES: THE LENGTH AND DIFFICULTY OF CONFESSING WRONGDOINGS IS THE LENGTH AND DIFFICULTY OF BEING IN THE MOUNTAIN STAGE OF THE TOUR DE FRANCE

We will use that model in our analysis of how hotels pictured by Condé Nast Traveller as places of rest for humans and points of departure to go out into the open to contemplate nature are metaphorically construed at various levels shown above. The topical transition from EGO to ECO is seen to be particularly manifested at the mental-space level, that is, at the level where meanings acquire their full actualization potential in a specific context. Wherever possible, we will establish conceptual pathway links between the rich contextualized meanings contained in these descriptions and their more schematic (decontextualized) conceptualizations.

Last but not least, Kövecses (2020: 165) distinguishes four types of the context where the denotational meaning is modulated:

1. Situational context: physical situation, cultural situation, social situation
2. Discourse context: surrounding context (co-text); previous discourse; knowledge about speaker/hearer, dominant forms of discourse
3. Bodily context: correlation in experience, bodily conditions, bodily specificities
4. Conceptual-cognitive context: metaphorical conceptual system, ideology, concerns and interests, history

Indeed, as the analysis below shows, not all types of context are equally activated in the subject-matter hotel descriptions, given the research focus of this paper. Still, we feel that the contextual variation is systematic and very much illustrative of the marked shift from chauvinist (self-interested) anthropocentrism to perspectival anthropocentrism, which sees EGO's viewpoint as inherently constrained by its bodily specificities, yet it claims EGO's openness to participate in the richness and wealth offered by ECO.

The following analysis of textual extracts from Condé Nast Traveller combines the two outstanding theoretical frameworks: at the form-level considerations, it draws on Bakhtin's (1986) reflections on speech genres, whereas at the content-level consideration, the chapter relies on the aforesaid ECMT (Kövecses 2020). Both approaches complement, thus offering an objectified perspective on the entire EGO-TO-ECO transition process outlined above. Last but not least, we also do not omit to take into consideration the aspect of reader responses to the analysed blurbs, as such feedback well marks up the ECO-sphere of the engaged form-content-reader interaction.

PART TWO

Our data

Condé Nast Traveller is a monthly magazine on everything to do with travel – hotels, restaurants, useful information about destinations around the world. A big part of each issue is taken up by hotel 'blurbs', descriptions varying from a sentence up to 300 words. These short texts are very evocative, both in terms of imagery and aimed emotional reaction. At first glance, this seems to be achieved by the use of complex nouns, adjectives and adverb phrases. At the same time, each text maintains the balance between emotive language and informative content and ends with practical information about the hotel, such as its website or the price per night.

This chapter is a continuation of the previous PALA[6] 2019 (Drewniok) and PALA 2021 (Drewniok and Kuźniak) papers. For the purpose of the analysis here, we have looked at a sample of *Condé Nast Traveller* issues (three issues from 2019 and three from 2021). Each issue contains at least one list, and we focus on the short hotel blurbs only.

1. April: 2019 The Editors' List 2019: our favourite affordable new hotels (fifteen texts)
2. May 2019: The Hotlist 2019. Our pick of the best new hotels in the world (fifty-two texts)
3. June 2019: The round-up Greece: the coolest new island hideouts (five texts)
4. April 2021: The round-up: Farm Stays (eight texts)
5. May 2021: The round-up: British Houses (seven texts)
6. June 2021: The Hotlist 2021 (Emerging into a new world of travel) – (forty-one texts in print – more on the web)

The form of these short texts can be examined on two contextual levels: the macrolevel would involve genre considerations and emerging themes (frame-level metaphorization), whereas the microlevel would be text structure and of special interest – lexical and grammar choices (domain-level and mental-space level metaphorization, see Kövecses 2020: 151).

Macrolevel: what genre is it?

Drawing on Bakhtin's (1986: 62) primary vs. secondary genre dichotomy, it can be said that these hotel blurbs are of secondary genre due to their complexity and multiple functions – they are to endorse a specific hotel, but also more generally to sell the recipient a specific perspective on and way of travel. We refer to them as 'blurbs' here for the lack of a better label, although they are very different to book blurbs. They include visuals (photographs) just like adverts or longer travel pieces/travelogues. Although they are persuasive pieces to entice the reader, they are positioned differently to the explicitly sponsored texts in the magazine (these are longer and clearly labelled as sponsored). As a secondary (complex) speech genre, they have absorbed primary genres mentioned above – adverts, travelogues or even hotel catalogues. The examination of this small *Condé Nast* corpus suggests that these hotel blurbs might be a new, separate, maybe even hybrid, category. At the level of frame conceptualization, the category of HOTEL BLURBS may be compared to MULTIFACETED INFOTAINMENT CHUNKS, where each facet serves a specific communicative end of persuasion. The structure is, as said above, hybrid in the sense that FACETS grounded in symbolic or iconic cognitive-conceptual context make up a uniquely identifiable category. At mental space-level representation it seems that HOTEL BLURBS come up as a result of the blend of the two input spaces, that is, ADVERTS and CATALOGUES, with the generic space close to what we conceive of as TRAVELOGUES.

The persuasive style in which these texts are written is consistent with advice that can be found online on how to write a catchy hotel description, for example, White Sky (2019):

1. Be accurate;
2. Start with the customer in mind;
3. Promise authentic hospitality;
4. Provide inspiration;
5. Write precisely and succinctly;
6. Provide vivid descriptions for images;
7. Use sensory language and phrases;
8. Make it about the experience;
9. Use an outline.

The only point that is not fully followed is writing precisely and succinctly – our texts contain complex phrases (especially adjective-adverb patterns) and long, complex sentences. This also goes against travel writing advice from Lee (2009: 92): 'Don't write as if you were writing a postcard'. Since 'genres always reflect the communicative purposes they are thought to fulfil' (Berkenkotter and Luginbühl 2014: 286), and the hotel blurbs examined here are not typical travel-brochure hotel descriptions, they could be seen as a type of postcard, saying 'wish you were here'. Because of their style, they close the gap, the distance, between the writer and the reader. This is supported by often addressing the reader directly – for example: 'come for fish pie', 'spend your days feeding the Hebridean sheep or walking … where you can fly above the patchwork quilt of land and sea' or 'don't expect … '; as well as the use of colloquialisms and neologisms which will be discussed later below.

Macrolevel: The themes

As already mentioned at the beginning, these short texts describe specific hotels or other similar hospitality spaces – it is done from the perspective of someone who has stayed there, using evocative language leading to some narrative transportation. In both the 2019 and 2021 texts, several general themes emerge, including

1. Visual language
2. Language of sounds
3. Food descriptions
4. Cultural inspirations

Firstly, there are many visual descriptions of the scene (both hotels and its surroundings), creating a mental image for the reader. This includes colours, patterns, design features and textures, for example: 'The skies … are streaked in red', 'towering black-and-white pillars', 'thick walls are tactile with bare plaster', 'in-built stone benches and bed platforms are softened with muted mustard throws'. Note how these phrases not only describe what a hotel guest would see, but also how they would experience their surroundings, for example, with towering pillars – implying their significant height, and by extension – the height of the space, or stone benches – suggesting a solid seat, especially when coupled with 'softened'.

Apart from the visual language, aural descriptions can be found in our texts, too. Some are to do with the music in the hotel bars or restaurants, but more often these refer to the hotels' surroundings, for example: 'the soundtrack is of the river whooshing by in the valley below' or 'fresh spring water gurgling from its fountain'. Note that both verbs are onomatopoeic here, contributing to the evocative effect. Hotel menu, chef's inspirations and where the food is sourced are often covered in the hotel blurbs examined here. The 2019 descriptions seem to aim to impress the reader – note the choice of verbs and overall structures: 'Fish pie studded with salmon and creamy scallops, or lobster hauled in from neighbouring Nelson Bay'; 'A short British menu hints at Handling's Gleneagles training, but there's playfulness alongside the veal sweetbread and Highland wagyu: alt-J on the sound system, and a Jammie Dodger with the strawberry-infused Old Fashioned.'

In contrast, the 2021 food content in the blurbs has much more ECO focus: 'Before dinner, the chef talks through the inspiration for the food and the source of ingredients. Once a week the wood-fired outdoor oven crackles for pizza night, when everyone grabs slices and eats while standing and chatting in the courtyard.' Such descriptions as above really illustrate the 2021 farm-to-fork perspective.

Many of the hotels included in the *Condé Nast Traveller* boast cultural inspirations from around the world, often with eclectic results. The 2019 text on the Bulgari Hotel in Shanghai emphasizes the mix between the brand's Italian heritage and the culture of the hotel's location: 'Inside, an immaculately choreographed tussle between Italian and Chinese design cues is enacted in marble and bronze versus silk and lacquer.' The two cultures are presented here as clashing by design – although they 'tussle', it is 'immaculately choreographed'. The materials mentioned – marble, bronze, silk and lacquer – are also set in opposition to each other. Perhaps the text suggests that such design choices aim to satisfy both guests familiar with Bulgari and those who want to experience more Chinese culture.

Interestingly, in the 2021 description of the Cornish Ukiyo, one can find a seemingly similar approach with a hotel mixing its cultural inspirations: 'the building's part Scandinavian, Californian and Japanese charm can claim your full attention.' However, this time all three styles are meant to be synonymous – minimalistic and laid-back. These evocative descriptions are achieved on the microlevel by a specific choice of lexis and grammatical structures, which will now be discussed in the following sections. All in all, given Kövecses's (2020) classification of contexts quoted above, the primary grounding of HOTEL BLURBS is situational on account of its focus on the physical situation, cultural and social situation in which the guest, seen as an engaged and mindful participant, is placed. This situation may be thematically framed as JOURNEY FROM EGO TO ECO CONTEMPLATION.

Microlevel: Lexis

One of the ways *Condé Nast Traveller* hotel blurbs evoke certain imagery is through hyphenated compound nouns. Some of them are common: 'low-key' or 'all-you-can-eat'. But there are many more unusual ones, especially when in a longer phrase with another noun or even a more complex phrase, for example: 'copper-sand beaches', 'raw-concrete architecture', 'eat-drink-sleep address', 'kick-your-shoes-off feeling staying at a friend's home', 'mud-under-the-fingernails ethos', 'juniper – and wild-sage-scented road'. These phrases contribute to the sentences being more complex (more of which in the section below) and thus requiring more cognitive effort from the reader. Consider 'raw-concrete architecture': first, the reader needs to retrieve the meaning of 'raw concrete' where 'raw' does not denote literally 'freshly poured/not yet hardened', but rather a more metaphorical meaning of untreated, unclad or unpainted concrete texture. Once this meaning is established, the reader then needs to activate a schema for a certain type of architecture. It is only after these two steps are actuated a specific image evoked by the hotel blurb comes up. On a domain-level of metaphorization, HOTEL BLURBS resemble COMMUNICATION AT GRASSROOTS, which derives from the schema NEAR IS BETTER (see Krzeszowski 1997).

Another way of creating specific travel imagery for the reader is using borrowings from other languages. However, these are only those that are commonly used and easily recognizable – at the same time bringing a flavour of a different culture, but not too foreign as to confuse the reader. The above-quoted schema of NEARNESS is thus still at work, though it is counterbalanced by the 'tamed' effect of the foreign. Examples include *Le talk* of Paris, cosy *deshabillé*, pied-à-terre hotel, pièce de resistance, agriturismo.

Apart from compounds and borrowings, the lexis of these hotel blurbs also includes colloquialisms, neologisms and even some ellipsis. Some colloquialisms are so common they do not really stand out ('sunnies' or 'bolthole'), but some are more marked – for example 'from kook to dude'. The use of colloquial language extends to whole sentences at times: 'The look could have been storyboarded for a rom-com' – note the verbalization of the storyboard here. It is no wonder then that neologisms are also present in the examined texts – some, once again, quite simple ('buzzy beaches' or 'clubby library'), but some are much more creative on both lexical and grammatical levels, for example: 'seemingly simple design unpacks a punch'. Like the hyphenated compounds mentioned above, this phrase also requires at least a two-step cognitive effort: first, to recognize 'packs a punch', and then to consider how 'unpacks' changes the meaning here. Again, NEAR IS BETTER schema is in focus, but NEARNESS does not entail closure in the contemplating self of the guest but embraces aspects of the situation context in which the guest is found. Somewhat imperceptibly, EGO-focused contemplation changes into ECO-oriented experience.

And finally, ellipses are common in our texts. They vary from short phrases such as 'a crisp white' [wine] or 'local brews' [beers], to longer phrases, for example: 'consistent swell [of waves] has nurtured a laid-back scene'. With such ellipses, the reader will need to fill in the gaps to access the meaning; sometimes, this calls for recognizing a phrase widely used ('crisp white'), and sometimes – selecting one meaning of several: in British English, the most common meaning of a 'brew' is 'tea', therefore a British reader will need to take a cognitive step to identify 'local brews' as beers and not types of tea. The ellipses provide for the effect of NEARNESS at the level of discourse context (see Kövecses 2020: 165).

Microlevel: Grammar

As can be expected, thanks to the use of many complex and compound noun phrases, the sentences in *Condé Nast Traveller* hotel blurbs can be significantly longer. Sometimes, they still are a simple sentence with one verb, but with a complex subject, for example: 'The hotel's overwhelming sense of calm, its welcoming staff and its nourishing farm-to-table menu – DYI poke bowls; spaghetti made with coconut milk; and salads of little-known local produce such as snake gourd and the herb *gotu kola* – are drawing in intrepid design-keen travellers.' Seen in cognitive-conceptual context, the passage conveys at the mental-space level the meaning of EXPERIENCING THE BLEND OF THE EXOTIC AND THE DOMESTICATED IN TRANQUILLITY as an optimal form of leisure-time activity. Note three separate subjects – 'sense of calm', 'staff' and 'menu' (which is an even more extended subject phrase), but only one verb. The use of the Present Continuous tense is conspicuous – it emphasizes the currency of the description. This long, descriptive sentence constitutes a larger part of the whole blurb. Apart from the noun-heavy grammatical structures, there are also many complex sentences featuring more than two verbs, for example:

> There is no-key policy, paired with encouragement to explore far-reaching grounds. The small concierge team build bespoke itineraries; visitors can spend time in the stables with La Donaira's very own cowboy Seamus Gaffney, embark on a farm tour to learn more about biodynamics, take part in guided stargazing, or pull a book from the thoughtfully curated shelf and unfurl in the spa.

At mental-space level of metaphorization, the above passage again evokes the meaning of PROXIMITY IS THE GUEST'S BEST COMPANION. Although not all the above verbs are action ones, the sheer number of them is marked – this is clearly a list of activities guests can enjoy, with the majority of verbs denoting 'experience': 'build', 'spend', 'embark', 'learn', 'take part' and 'unfurl'. Most of the verbs here also require an object – and they take here more than a one-word NP or adverbial. The two above examples also illustrate well the change of focus in the blurbs – in 2019, the focus was more on the individual experience (EGO), whereas, in 2021, the emphasis is more laid on the experience of the particular location (ECO).

The shift from EGO to ECO

Our analysis above shows that there was a visible shift in *Condé Nast Traveller* hotel blurbs between 2019 and 2021. The language to deliver the content – the lexis, the grammar and the overall style – has not changed. Although these blurbs are written by a team and not one single person, the style is consistent and recognizable. However, the content of these blurbs changed.

In 2019 the focus was on the individual experience of the traveller – often shown as luxury hotels, exceptional service, links to art and exquisite food. The locations presented were spread around the globe. In contrast, in 2021, the focus shifted from the individual to being part of something bigger, of a local ecosystem. This included many descriptions of nature, farm-to-fork dining and overall sustainability. The locations chosen were closer to home – new, smaller hotels (hideaways), farmhouses or stately British homes. The recent pandemic must have contributed to this shift, but this ECO-focus started much earlier, as we discussed at the beginning of this chapter. In the short term, when the world was still opening up in 2021, a focus on travel closer to home and sustainability is expected. However, this ECO-direction continues to this day – despite locations further afield being explored once again in 2023 issues.

The recipient

When looking at these texts, there is an important question to be asked: who is the target recipient of these hotel blurbs (and the whole magazine for that matter)? According to *Condé Nast Traveller (UK)* media kit, they have someone specific in mind, and it is not just an ideal reader, but data based on the actual readership (although the target reader of the media kit is advertisers, so we can wonder if it is really actual or just the target).

The Condé Nast Britain[7] website states:

Condé Nast Traveller, launched in 1997, has been at the vanguard of a new age of exploration and luxury travel. Today, Condé Nast Traveller's influence extends to multiple platforms including the website, social media, books and a series of Condé Nast Traveller events. Targeting an audience of the influential and curious, Condé Nast Traveller is the ultimate luxury travel brand.

While reading the passage above, the recipient is conceptualized as 'influential and curious'. Both adjectives encourage the formation of the mental space of an ENGAGED AND MINDFUL GUEST, empowered to actively cooperate with their environment in the contemplation of ECO. The emerging LUXURY is no longer contextualized as a sole emanation of materialism paired with overabundant consumption. It is rather seen as a welcome outcome of complex bodily-spiritual experience, where it is ECO that defines EGO, not vice versa.

Not only does the magazine present itself as a specialist one in luxury travel but also describes its target audience as influential (powerful rather than rich?) and curious ('anti-tourist?'). And when you delve into the actual media kit, it says – based on their analytics – that the target reader spends over £9k per year on travel and over £8k on fashion. Interestingly (but not surprisingly), they put their readership in ABC1 social grade. We have decided to refer here to the target recipient rather than the target reader (or subscriber), because there might be multiple recipients in one household (with one copy of the magazine) or they might pick up a copy somewhere else. The 2021 media kit[8] emphasized the shift in luxury tourism, in the words of the then Editor-in-Chief Melinda Stevens:

> Condé Nast Traveller celebrates and supports the most creative and sophisticated adventures and endeavours around the globe, shining a light on curious places, people and trends and encouraging readers to see the world afresh – not just with relish but with respect. In 2021, its travel advice and access will prove more valuable in shifting sands environment providing trusted, discerning, first-hand, up-to-the-minute, inside-track reports on destinations, sustainability, culture, wellness, and food, with high-low mix and evocative, transportive storytelling alongside the world's best travel photography.

This approach continues to be reflected in the 2022 media kit[9]:

> Condé Nast Traveller is a leading and opinionated voice on travel and culture, driving conversations around sustainability, inclusivity and transformation, showcasing the local and authentic, and a force for real connection and change.

Divia Thani, Global Editorial Director:

> The focus for Condé Nast Traveller moving into 2022 is one of togetherness, the global made local. Spearheaded by Global Editorial Director Divia Thani, the lens will be focused on people in places – the stories, experiences, craft and creativity of a destination's people are what bring it to life. People will be at the centre of our global print editions and digital platforms.

This supports our findings from the content of 2021 blurbs, where there is a shift from ECO (focus on individual luxury experience) to ECO (sustainability, respect). The 2021 media blurb refers explicitly to what they're trying to achieve in their texts – 'evocative, transportive storytelling', and this is clearly a long-term goal as reflected in the 2022 media blurb which mentions sustainability, transformation, force for … change, the global made local.

Based on our findings on the content of the hotel blurbs examined here and on how *Condé Nast Traveller* brands itself, it is quite clear that the target recipient is seen as more than just a tourist – it is someone curious, respectful of their surroundings, caring for the environment and the planet, and not just consuming like a stereotypical tourist. This idea was explored by McWha et al. (2016), who focused on longer travel articles in specialist travel magazines, including *Condé Nast*, and where they labelled such a more conscious traveller as an anti-tourist. 'The discourse of authenticity seems to have been adopted by these travel articles to maintain the anti-tourist identity' (McWha et al. 2016: 94). 'The tendency in the articles to highlight responsible tourism further reinforces appeals to the anti-tourist, where the reader could be persuaded that they can become a traveller by engaging in responsible tourism' (McWha et al. 2016: 97). Although the hotel blurbs are not the same as longer travel articles, this could apply to them as well.

Apart from the target recipient, it is also important to consider any actual recipients who might be more aspirational travellers than seasoned in luxury travel. They read for pleasure and escapism. Unlike the target recipient, who draws on the luxury travel schema from their lived experience, the aspirational reader experiences a deictic shift in a way – imagining a perspective of a luxury traveller and activating a schema for this imagined reader, not themselves. They are also perhaps experiencing pseudo-nostalgia 'which links nostalgic feeling to other forms of escapism such as fantasy or day-dreaming and even to future expectations' (Hunt and Jones 2013: 14).

Conclusion

To conclude, in this short project we have looked at a snapshot of hotel blurbs included in *Condé Nast Traveller* (UK). Although we have only looked at six issues – three from 2019 and the same three months in 2021, they are representative of the overall style of the magazine, which includes such texts and various lists in every issue. We claim that these blurbs are their own separate and relatively new category, and as such – they have their own style. Comparing 2019 and 2021 there is little change on the microlevel – text style, structure and length, as well as lexical and grammatical choices. There are also the same overall general themes reflected. However, on a more macrolevel, there is a visible shift of focus and the choice of destination (both place and hotel), most likely affected by the pandemic and restrictions to travel. But there is also another shift in luxury travel – visible in the texts and also explicitly mentioned in the media kit descriptions: this is a shift from EGO (focus on one's own luxury travel experience, being a consumer) to ECO (being part of the ecosystem, conscious and responsible participant) at various levels of conceptual metaphorization with the situational context (physical, cultural, social embedding) as a trigger of the entire process. This shift is not necessarily affected by the pandemic. Future explorations could include looking more closely at when exactly this shift started happening.

Notes

1 Translated with DeepL.com (the original passage in Polish).
2 There are two main ways of understanding anthropocentrism. One view equates it with human chauvinism (chauvinistic anthropocentrism), while the other treats it as the inability of humans to transcend the human perspective (perspectival anthropocentrism). Both of these meanings are important in the analysis of the critique of anthropocentrism carried out in deep ecology (Hoły-Łuczaj 2018: 51).
3 The term 'deep ecology' first appeared in 1973, in Naess's article The Shallow and the Deep, Long-Range Ecology Movement. Here Naess outlined the project of 'deep' ecology, contrasting it with the dominant, 'shallow' ecology movement. As such, shallow ecology supports solutions, such as recycling, that would address the deepening ecological crisis. In doing so, however, shallow ecology, as Naess points out, avoids questions about the source of the ecological crisis. It deals only with ad hoc attempts to address its symptoms. In Naess's view, on the other hand, a way out of the ecological crisis will only be possible when the worldview that caused it is challenged. This is precisely the task that deep ecology sets itself. The change that deep ecology seeks, therefore, is not so much a reform of certain human behaviours as a radical transformation of the attitude that humans take toward the world around them (Holy Łuczaj 2018: 25–26).
4 According to this position, it is impossible to create a hierarchy within the various forms of broadly defined 'life', identical to the realm of physis (plants, animals, rocks, rivers). According to this assumption, every (natural) entity has an 'equal right to live and flourish' and equal inherent value, constituting an end in itself (Hoły-Łuczaj 2018: 9).
5 As Hoły-Łuczaj (2018: 49) emphasizes, in deep ecology, when a choice has to be made between the life needs of two entities, always, it should be emphasized,

priority will be given to beings close to us and known over distant and directly unknown ones. At the same time, however, life needs must not be confused with trivial interest. This is explained by Bill Devall (1988), who says that the needs of life are biological, social, as well as spiritual. There are also many ways to fulfil these needs. For example, one may have a life need to relax in nature, but that does not mean that our life need is to ride an ATV through a gentle meadow. Life's needs, Devall (1988) adds, should be met by the simplest, most elegant and least destructive means possible to the environment.

6   Poetics and Linguistics Association (PALA) annual conference.
7   Condé Nast Britain https://www.condenast.co.uk/cn-traveller/ (last accessed 20 February 2023). Note that this text has not changed since 2019.
8   Archived by the Wayback Machine 20 July 2021: https://web.archive.org/web/20210720142907/https://cnda.condenast.co.uk/static/mediapack/tr_media_pack_latest.pdf (accessed 20 February 2023).
9   Available at: https://cnda.condenast.co.uk/static/mediapack/tr_media_pack_latest.pdf (accessed 21 February 2023). As of February 2022, the new 2023 media pack is not yet available.

References

Bakhtin, M. M. (1986), *Speech Genres and Other Late Essays*, Translated by V. W. McGee, edited by C. Emerson and M. Holquist, Austin: University of Texas Press.

Berkentkotter, C. and M. Luginbühl (2014), 'Producing genres: Pattern variation and genre development', in E. M. Jakobs and D. Perrin (eds), *Handbook of Writing and Text Production*, Berlin: De Gruyter Mouton.

Bogusławska, M. (2022), 'Introduction. ecolinguistics in the New Millenium (noted in the year 2022)', in M. Bogusławska, A. A. Dragoescu Urlica and L. Kamberi (eds), *From Cognitivism to Ecologism in Language Studies*, 9–16. Berlin: Peter Lang.

Bogusławska, M., Dragoescu Urlica and A. A Kamberi, L. eds (2022), *From Cognitivism to Ecologism in Language Studies*, Berlin: Peter Lang.

Devall, B. (1988), *Simple in Means, Rich in Ends: Practicing Deep Ecology*, Salt Lake City: Gibbs Smith.

Evans, V. and M. Green (2006), *Cognitive Linguistics. An Introduction*, Edinburgh: Edinburgh University Press.

Hoły-Łuczaj, M. (2018), *Radykalny nonantropocentryzm. Martin Heidegger i ekologia głęboka*, Warszawa: Wydawnictwo Uniwersytetu Warszawskiego.

Hunt, L. and N. Jones (2013), 'Image, place and nostalgia in hospitality branding and marketing'. *Worldwide Hospitality and Tourism Themes*, 5 (1): 14–26.

https://www.deepl.com/translator (accessed 10 January 2023).

https://www.condenast.co.uk/cn-traveller/ (last accessed 20 February 2023).

https://web.archive.org/web/20210720142907/https://cnda.condenast.co.uk/static/mediapack/tr_media_pack_latest.pdf (accessed 20 February 2023).

https://cnda.condenast.co.uk/static/mediapack/tr_media_pack_latest.pdf (accessed 21 February 2023).

Johnson, M. (1987), *The Body in the Mind: The Bodily Basis of Cognition*, Chicago: University of Chicago Press.

Kardela, H. (2006), 'Metodologia językonzawstwa kognitywnego', in. P. Stalmaszczyk (ed.), *Metodologie językoznawstwa. Podstawy teoretyczne*, 196–233. Łódź: Wydawnictwo Uniwersytetu Łódzkiego.

Kövecses, Z. (2006), *Language, Mind, and Culture: A Practical Introduction*, Oxford: Oxford University Press.

Kövecses, Z. (2015), *Where Metaphors Come From: Reconsidering Context in Metaphor*, Oxford: Oxford University Press.

Kövecses, Z. (2017), 'Levels of metaphor', *Cognitive Linguistics*, 28-2: 321–47.

Kövecses, Z. (2020), *Extended Conceptual Metaphor Theory*, Cambridge: Cambridge University Press.

Krzeszowski, T. P. (1997), *Angels and Devils in Hell. Elements of Axiology in Semantics*, Warszawa: Wydawnictwo Energeia.

Kuźniak, M. (2021), *The Geometry of Choice: Language, Culture, and Education*, Cham: Palgrave Macmillan.

Lakoff, G. (1987), *Women, Fire and Dangerous Things: What Categories Reveal About the Mind*, Chicago: University of Chicago Press.

Lakoff, G. and M. Johnson (1999), *Philosophy in the Flesh*, New York: Basic Books.

Lee, C. (2009), *Word Bytes*, Carton: Melbourne University Press.

Lovejoy, A. O. (1971), *The Great Chain of Being: A Study of the History of an Idea*, Harvard: Harvard University Press.

McWha, M. R., W. F. Laing Frost and G. Best (2016), 'Writing for the anti-tourist? Imagining the contemporary travel magazine readers as an authentic experience seeker', *Current Issues in Tourism*, 19 (1): 85–99.

Naess, A. (1973), 'The shallow and the deep, long-range ecology movement', *Inquiry*, 16: 95–100.

Naess, A. (1989), *Ecology, Community, and Lifestyle. Outline of an Ecosophy*, translated and edited by David Rothenberg, Cambridge: Cambridge University Press.

White Sky (2019), '9 essentials of a catchy hotel description', Available online: https://whiteskyhospitality.co.uk/9-essentials-of-a-catchy-hotel-description/ (accessed 7 February 2023).

2

A Diffractive Approach to Reader Response (with Reference to Barnes's *The Sense of an Ending*)

Amélie Doche

Introduction: From ego to eco

This chapter aims to challenge traditional deductive reader-response research, marked by an all-knowing researcher assuming causal links between literary text and response. To do so, I lay the foundations for an inductive and 'diffractive' reader-response, which rejects the binary oppositions between object and subject, research and researcher, and text and response.

In the long-standing Western tradition, conceiving of oneself as being 'in the world' paves the way for a representational and ego-centric ontology nourished by the Cartesian's subject-object divide (Descartes [1941] 1993). Heidegger's ([1927] 1962) idea of *Dasein* challenges this world view by establishing 'being-in-the-world' as a relational state signalling a constitutive relationship between everyone and everything. Haraway (2016: 14) goes further by stating that we are '*of* the world' (her emphasis). Heidegger's and Haraway's thoughts inspire a relational and eco-centric ontology, that is, an ontology privileging mutual relations over distinct entities. The critique of a world view based on the opposition between subject and object is not new: it goes back to the Romantic scientist Goethe according to whom the known (i.e. the object of analysis) always alters the knower (i.e. the scientific observer). According to this conception, to touch is to be touched, and to read is to be read.

The complex interrelations between texts and readers lie at the heart of response-oriented stylistics (see Mason 2019; Peplow and Whiteley 2021; Doche and Ross 2022), which – despite emphasizing the dialogic relationship between text and response – assumes a clear distinction between known and knower. Two recent exceptions to this long-established Western paradigm are

Escott (2021: 197) and O'Halloran's (2023) argument that literary practices be considered 'assemblages': our visible interactions with texts are embedded in other less visible interactions with people, places, objects and media. No one can escape these extra-textual entanglements which are never 'extra' as far as they condition the textual. The American physicist and philosopher Barad advances that the false dichotomy between knower and known embodies itself in the term 'interaction'. While the latter does not preclude mutually modifying relations, Barad (2007: 33) argues that being-of-the-world lends itself to intra-actions, defined as 'the mutual constitution of entangled agencies'. Throughout this chapter, I privilege the term intra-action for two reasons: a) I consider that online reviewers are engaged in various relationships which might shape their responses to *The Sense of an Ending* (henceforth *TSOAE*) and b) as a (re)reader and researcher of *TSOAE*, I cannot observe readers' responses from afar.

By considering that other variables might shape readers' responses, this chapter challenges the base assumption in reader-response stylistics that the stylistic features displayed in readers' reviews respond to the style of the book reviewed (null hypothesis, H0: literary style and literary response are entangled). Because of this assumption, confounding variables/intra-actions are left out. Until the recent work of Escott (2021) and Stockwell (2021), researchers officially remained 'outside' of the analysis, thus obliterating a major variable: the researcher's relationship with the data – also known as the observer's paradox – which defines *a priori* which analytical framework is used. By considering a one-to-many relationship (text, reader, researcher and medium), this chapter aims to make confounding variables visible. Thus, my alternative hypothesis is that the language of the reviews is shaped by many intra-actions (H1: literary response is entangled with many variables). This chapter aims to answer the following questions:

1. What makes readers' responses to *TSOAE* unique, by which I mean specific to the readers' intra-actions with the novel?
2. By implication, what makes readers' responses to *TSOAE* different from readers' intra-actions with the online book reviewing genre?

Thinking diffractively

Influenced by the work of feminist philosopher and scientist Barad (2007), this reader-response study uses the theoretical concept of 'diffraction' to approach the one-to-many relationships involved in reader-response. In physics, diffraction 'has to do with the way waves combine when they overlap and the apparent bending and spreading of waves that occurs when waves encounter an obstruction' (Barad 2007: 74).

The concept of diffraction – widely used in feminist and new materialist philosophies – has been applied in other fields, such as literary translation (Doche 2021), higher education (Baker 2021) and contemporary art (Sayal-Bennett 2018). Here, I am modelling what a diffractive approach might look like in reader-response stylistics. I am also hoping to show how diffraction helps neutralize confounding variables. Central to the phenomenon of diffraction is the notion of difference – when waves encounter an obstacle, they change shapes. Varying obstacles create varying shapes or 'diffraction patterns' (Figures 2.1 and 2.2). Rather than simply observing differences and similarities between literary style and literary response, diffraction forces me to rethink the nature of causality: which intra-action causes X stylistic feature in the reviews? I start with the effect – that is, the language features displayed in the reviews – and work backwards to identify the cause. Following Barad (2007) and Escott (2021), I assume that online reviews intra-act with several factors: the online socio-material environment (here, Amazon), the offline socio-material environment and genre expectations within the discourse community, to name but a few. Since the many extratextual factors conditioning the reviews remain out of

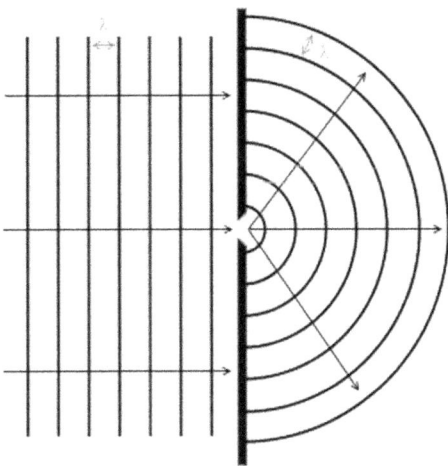

Figure 2.1 Propagation of a wave through a chink. 2008. This file is licensed under CC BY 4.0.

Figure 2.2 Diffraction in ocean waves. 2009. © Exploratorium. This file is licensed under CC BY-NC-SA 2.0.

reach, my research only considers two possible intra-actions: the novel *TSOAE* and book reviewing practices on Amazon.

I focus on one positive and one negative review posted on Amazon. I label those R1 and R2 referring both to the reviewers and their reviews of *TSOAE*. On the date of data collection – 9 November 2019 – the first review posted by R1 awarded *TSOAE* five stars and features as 'the top review for the United Kingdom' on Amazon. The second review, posted by R2, is the 'most helpful' two star-review. At the time of writing – 21 November 2022 – R1 is no longer the top review and R2 is no longer the most helpful two-star review. I compare each of the two reviews with two other reviews produced by the same reviewers to account for the reviewers' relationships with genre expectations on Amazon. I settled on two reviews for the simple reason that my diffractive approach required me to read the two other books reviewed by R1 and R2 to objectively neutralize the reviewers' intra-actions with those books. Reading the novels reviewed and examining three reviews written by the same reviewer enables me to see the instances where the reviewer's style responds to *TSOAE* and the instances where it responds to the online book reviewing genre, here understood as the 'rhetorical conventions and stylistic practices that are tacit and routine for the members of the [Amazon] discourse community' (Doheny-Farina 1992: 296). I neutralize the review's intra-action with genre conventions in the online socio-material environment – one of the many entanglements within which the reviews occur – to show the specific intra-action between *TSOAE* and the reviewers. This approach enables me to extract a variable from a constant, that is, to identify a difference from repetition in the space of the review.

As a reader and researcher, I am myself entangled with the phenomenon which I seek to explain. I attentively researched *TSOAE* for my MRes thesis defended in June 2020. My thesis comprises a dialogically oriented stylistic analysis of *TSOAE* followed by an Appraisal analysis of three Amazon reviews (Doche 2020). My reading of *TSOAE* entails that some of the features deployed in the reviews will seem particularly 'attractive' (figure) to me. Similarly, some features will naturally be 'neglected' (ground). Therefore, I focus on the passages in readers' reviews that most attract my attention because of a perceived familiarity with Barnes's novel. Diffractive research rejects pre-existing frameworks for the simple reason that these frameworks separate a priori the data that will come to matter from the data that will not ('matter' in both senses of the word, that is, to be significant and to be materialized). Let us, however, not fall under the illusion that diffraction can be methodology-free. While I analyze the reviews using the stylistic tools which lend themselves to the analysis, my knowledge is limited, and so are the tools I know. Therefore, it is necessary for me to disclose that I am familiar with Text World Theory (Gavins 2007; hereafter TWT), Dialogic Syntax (Du Bois 2014; hereafter DS), Appraisal Theory (Martin and White 2005), French narratology and postmodern philosophy. When necessary, I elaborate on the relevant tools as I use them in the analysis.

R1's intra-action with *The Sense of an Ending*: Repetition and order-wor(l)ds

R1's five-star review of *TSOAE* – posted in 2018 – will be compared to two other reviews: *The Daylight Gate* (henceforth *TDG*) by Jeanette Winterson (five stars) and *The Fall* (henceforth *TF*) by Albert Camus (four stars). Jeanette Winterson and Julian Barnes are both British writers. While Albert Camus is a French writer, he is regularly taught at GCSE and A-Levels in the UK. This could suggest that most experienced readers have encountered Camus's work. Here, Julian Barnes, the reviewers and myself all belong to the same discourse-world: contemporary Britain. It is expected that, to some extent, Julian Barnes, R1 and myself share the same experiential knowledge.

Upon first reading, the following passage particularly attracted me because it seemed to echo the voice of the fictional character: '[t]he narrative constantly deconstructs itself, pathetically seeking corroboration of its flimsy acts of recall. Adrian's diary is a piece of the Webster jigsaw that can't be bequeathed, with all the legal wrangling in the world, because getting it "might disrupt the banal reiteration of memory"'. Although the reviewer quotes only one passage from the novel using quotation marks, the whole passage feels like *déjà-vu*. In fact, the reviewer uses the words 'pathetically', 'corroboration', 'bequeathed' and 'legal', all of which are uttered directly by Tony Webster in *TSOAE*. The phrase 'flimsy acts of recall' does not occur in the text. Yet, the novel opens with 'I remember, in no particular order' (Barnes 2012: 3) which functions as an 'empty world' in TWT, that is, as a minimal deictic space in which the 'function-advancing propositions must be inferred' by readers (Lahey 2004: 26). The exophoric reference 'I remember' points to something that is not accessible. While the reviewer includes Tony's voice in the subordinate adverbial clause of reason starting with 'it might', it seems particularly difficult to distinguish the framing voice in the real-world from the framed voice of the fictional universe.

In their review of *TF*, the reviewer also addresses the character's attitude towards the law: '[I]t concerns a Parisian lawyer, Jean Baptiste Clamence, who, in his advocacy work, turns the judiciary around to suit his own sense of the high moral ground. [...] As the confessional evolves, Clamence peels away the layers of his over-inflated ego and becomes a "judge-penitent" who finds a deadening immobility in himself, a catastrophic incapacity to act'. Interestingly, in both reviews, the reviewer uses attitudinal lexis: see 'pathetically', 'over-inflated' and 'catastrophic'. The difference between the two passages emerges from the different sources of attitudes: in the review of Barnes's novel, the reviewer concurs with Tony Webster's self-description as 'pathetic', which means that the negative attitude displayed emanates from a fictional character.

By contrast, in the review of Camus's novel, the negative attitude comes from the reviewer themself. Moreover, apart from 'judge-penitent', none of the content words used by the reviewers in the excerpt quoted above occur in *TF*. *TSOAE* and *TF* both feature a first-person homodiegetic narrator. However, the reviewer's response to these narrators varies: while they detach themselves from Clamance, they show strong engagement with Webster. Here, the interference between *TSOAE* and the reader has left textual marks which show where difference is being enacted. The text's intra-active influence has materialized a reduction in agency on the part of the reviewer, who seems to have first absorbed and then reproduced the voice of the fictional character. This observation recalls the point made by Stockwell: much literary criticism tends to re-enact the phenomenon under study 'in the rhetorical texture of the critic's language' (2009: 17).

The proximity between the reviewer (discourse-world entity) and Tony Webster (text-world entity) seems to be supplemented by proximity between the reviewer and the author, Julian Barnes (discourse-world entities). The review ends with the following paragraph:

> [I]f you feel let down by the sense of an ending on p.151, that's precisely, I suspect, what Barnes wants us to feel. Scratch your head into the early hours as you try to work it all out if you want, or, better, turn your watch over and think about it again while you're flipping over the fried eggs in the morning.

Firstly, the review shows an instance of 'hypothetical intentionalism' insofar as the reviewer makes hypotheses about authorial intentions (Rojek 2016). Secondly, the reviewer gives a recipe on how to read *TSOAE*. While my MRes thesis (Doche 2020: 63) argues that the reviewer's engagement with Barnes proceeded from the reader-reviewer's engagement with the narrator, reading the reviewer's other reviews proves me wrong: 'and if you feel let down by the sense of an ending on p.151, that's precisely, I suspect, what Barnes wants us to feel' (R1's review of Barnes's *TSAOE*); 'In the end, though, its meaning is murky and ill-defined, precisely, I suspect, what Camus desired' (R1's review of Camus's *TF*). These examples show that one of the reviewer's reviewing practices involves negotiating a sense of intimacy with authors. Rojek (2016) terms such practice 'presumed intimacy' insofar as there is no physical relationship between the reader and the author. R1's intra-actions with the online socio-material environment sustain frequent instances of 'presumed intimacy'. I have here illustrated this practice by presenting its strongest realization through parallelism: adverb 'precisely' + pronoun 'I' + mental verb 'suspect' + determiner 'what' + name of the author + desiderative verb. As the syntactic and semantic resonances suggest, 'presumed intimacy' should be treated as a recurrent pattern which is not contingent upon the reviewer's attitude towards *TSOAE*. Not only does the reviewer foster a sense of intimacy with spatiotemporally removed authors, they also establish bonds with imagined readers in the online community. The examples below give us insights into the reviewer's understanding of the book reviewing genre, which, for them, includes connecting with past and future readers alike. The reviewer first provides advice to future readers using the imperative form. Then, the reviewer establishes a connection with readers who have already finished the novels by alluding to some of the tangential themes developed throughout:

> Scratch your head into the early hours as you try to work it all out if you want, or, better, turn your watch over and think about it again while you're flipping over the fried eggs in the morning' (review of *TSOAE*).

> 'Read it before eating and it will spoil your appetite; read it after and buy a new rug! (review of *TDG*).

Focusing on differences in repetitions leads me to conclude that the reviewer's presumed intimacy with both authors and putative addressees on the online platform does not result from the intra-active influence of the novel.

As suggested earlier, the reviewer absorbs the voice of the fictional character. This observation does not match the reviewer's extensive use of quotations, which indicate intertextual references (and thus a plurality of selves and voices). Although the reviewer's reviews of *TF* and *TDG* do not feature any intertextual references which would open their dialogic scope, the reviewer introduces difference by positioning themselves as the source of appraisal. In their review of Barnes's novel, the reviewer does not accept their position as 'other'; rather, intertextuality functions as an 'order-wor(l)d' (Deleuze and Guattari [1980] 1987: 75) which territorializes the textual world in the discourse-world. I gather the six intertextual references in Table 2.1.

Table 2.1 Intertextuality in R1's review of *The Sense of an Ending*

R1's intertextual references to *TSOAE*	Position in the clause
Julian Barnes is brilliant at these deflections and refractions of sense, 'where the imperfections of memory meet the inadequacies of documentation'.	Intertextual reference: subordinate adverbial clause. No framing element.
Adrian's diary is a piece of the Webster jigsaw that can't be bequeathed, with all the legal wrangling in the world, because getting it 'might disrupt the banal reiteration of memory. It might jump-start something – though I had no idea what.'	First intertextual reference: subordinate adverbial clause. Second intertextual reference: main clause. No framing elements.
Tony can't 'travel'; he is stuck in 'subjective time, the kind you wear on your wrist', where the watch has been flipped over.	Semicolon: joins two independent clauses. Intertextual reference: main clause. No framing element.
This is 'true time', according to Tony, which 'is measured in your relationship to your memory'.	Intertextual reference: relative subordinate clause. Framing element: 'according to'.
Julian Barnes celebrates the fluidity of the displaced and disruptive voice of a man who, according to Veronica, 'never gets it, and never will'	Intertextual reference: subordinate clause. Framing element: 'according to'.
'Time', the story spells out, 'is not a fixative – it's a solvent'.	Intertextual reference: main clause. Framing element: 'spells out'.

It is worth pointing out that, in the first three instances, the reviewer introduces the intertextual reference with no framing words. A TWT approach would argue that the appearance of direct speech triggers a deictic shift and signals a move into a different ontological domain. In *praxis*, the reviewer does distinguish the discourse-world from the text-world. Upon closer investigation, however, one can observe instances of (ontological) metalepsis, which Genette defines as 'any intrusion by the extradiegetic narrator or narratee into the diegetic universe [...] or the inverse' ([1972] 1980: 234–5). The first example displays a semantic resonance between the phrase 'deflections and refractions' in the main clause and the phrase 'imperfections and inadequacies' in the subordinate clause.

According to the *OED*, 'refraction' is a type of 'deflection' and 'imperfection' is a type of 'inadequacy'. In various online dictionaries (such as thesaurus.com and Word Reference), 'refraction' appears as a synonym for 'deflection', and 'imperfection' appears as a synonym for' inadequacy'. According to the same sources, 'deflection' is not a synonym for 'refraction' and 'inadequacy' is not a synonym for 'imperfection'. The semantic analogy between two distinct ontological levels creates an ontological confusion, which appears to be reinforced in the second intertextual reference: the reviewer reports Tony's words without changing the pronoun 'I' to 'he' in 'I had no idea what'. Thus, the reviewer seems to be 'comfortable inhabiting the projecting text-world persona' of Tony Webster (Gavins 2007: 86). Inhabiting the textual persona leads the reviewer to emulate Tony's attitude by putting the story's and Veronica's 'voices' at a distance – at least grammatically. In fact, Table 2.1 shows that the reviewer inserts a framing device (see 'according to' and 'the story spells out') when they refer to either the story or Veronica. In *TSOAE*, Tony Webster, the first-person homodiegetic (and unreliable) narrator, keeps putting 'the story' at a distance. Moreover, as a rejected lover, he does not accept Veronica's blame for 'not getting it'.

I briefly introduced the term 'order-wor(l)d' to approach the function of intertextuality in the review. The reviewer does not adopt a critical distance towards the novel; rather, they seem to have absorbed the words and events coloured through Tony's consciousness. The reviewer's intra-action with the novel has left textual marks, namely the solid mental engagement on the part of the discourse-world entity towards the text-world entity and ontological confusions. These marks show where the effects of the reviewer's relationship with *TSOAE* emerge. The reviewer repeats the textual world: no new information is added. In fact, to the reader of Barnes's *TSOAE* and the reader of Amazon reviews, this review feels like *déjà-vu*. This 'order-wor(l)d' fixes the reading experience: what can be heard is an echo resonating within (and, one could suggest, despite) the reviewer. The review foregrounds a sense of monoglossia, that is, a sense that Julian Barnes's, Tony's and the reviewer's voices are but one. The reviewer's interference from *TSOAE* has led them to re-enact the literary phenomenon.

R2's intra-action with *The Sense of an Ending*: Difference and pass-wor(l)ds

In 2011, R2 awarded *TSOAE* two stars. Their review will be put in resonance with two other reviews: *Never Let Me Go* (henceforth *NLMG*) by Kazuo Ishiguro (awarded four stars) and *Hothouse Flowers* (henceforth *HF*) by Lucinda Riley (published in the UK as *The Orchid House* and awarded five stars). Kazuo Ishiguro is a British writer; Lucinda Riley is an Irish writer who grew up in England. As such, the discourse world remains contemporary Britain. What particularly struck me when I first read R2's review was their repeated use of the phrase 'get it'. In Julian Barnes's *TSOAE*, Veronica Ford characterizes the homodiegetic narrator as someone who 'doesn't get it':

1. 'Well, what's the next line? **You don't get it, do you?**' she said.
2. '**You just don't get it, do you?** You never did, and you never will.'
3. Veronica's saying, '**You just don't get it** … You never did, and you never will.'
4. Her reply went: '**You still don't get it.** You never did, and you never will. So stop even trying.'
5. I could have used the phrase as an epitaph on a chunk of stone or marble: '**Tony Webster – He Never Got It.**'
6. 'I'm sorry. **I just didn't get it.**' I retired to my table and waited for my supper.

In *TSOAE*, the verb 'to get' is only used in its negative form when it realizes a mental meaning (Doche 2020: 46) – thus, the mental meaning tends to be foregrounded (Hidalgo-Downing 2003: 321). These observations lead me to suggest that the reader is encouraged to engage with the phrase 'don't get it'. While the review's title 'Didn't completely get it' would signal that the reviewer shares Tony's characterization, the reviewer positions themself at the other end of the scale, that is, as someone who does get it: 'I did "get it" regarding the time/water parallel, and I certainly got it regarding the fragments of history we choose to suppress, or keep, or throw away according to whether we feel guilt, or remorse, or nostalgia. How many of us would actually recall a letter, word for word, written 40 years ago?' Both the emphatic 'do' and the adverb 'certainly' emphasize positive polarity in that they dismiss possibilities for doubts. Polarity refers to a choice between positive and negative. In the novel, Veronica keeps telling Tony that he 'does not get it'. Thus, the negation 'don't get it' – by virtue of its repetition in the text – becomes foregrounded by parallelism (Du Bois 2014: 370). The negative polarity emphasized in the novel contrasts with the positive polarity deployed in the review.

Du Bois's DS describes such contrast as a focal resonance. The contrastive element, 'deployed against a framing background of parallelism', comes to the foreground (Du Bois 2014: 382). Here, focal resonance embodies Bakhtin's notion of dialogism: the reviewer responds to the narrator's voice in their own voice, creating a 'living, tension-filled interaction' which both builds on and strikes a dissonance with the initial utterance ([1981] 1986: 279). Meanwhile, in their review of *NLMG*, the reviewer also employs 'get' in its mental capacity, and within single quotation marks, the verb does not jut out: 'this is the beauty of the novel, which some people may not "get"'.

In *NLMG*, 'get' does not characterize the first person homodiegetic narrator, Katy H. Rather, the fictional characters who 'get it' or 'do not get it' are Tommy and Ruth. In the review of *TSOAE*, 'get it' marks an exophoric reference to the 'text-world', that is, the readers' mental representations of the diegesis. However, in the review of *NLMG*, 'get' functions as an anaphoric reference to the beginning of the review: 'I do have to comment on all the previous reviewers who gave this fewer stars than I did. Their gripes seem to be with all the unanswered questions – why so much was left unsaid. They are missing, I think, Ishiguro's point'.

In the reader's review of Barnes's *TSOAE*, the phrase 'get it' forms part of the diffraction pattern. It is worth emphasizing that Veronica's words are uttered through direct speech, which, from a reader's perspective, leads to stronger mental representations of text-world utterances than does indirect speech.

Tony Webster: 'never gets it' R2: 'certainly get it'
(Negative polarity) (Positive polarity)

Figure 2.3 Focal resonance between Tony Webster and Reviewer 2 (R2 Chapter 2).

The pronoun *you* may function as a double-deictic (Herman 1994) as far as it encompasses at least two deictic roles: one internal to the discourse event, that is, the text-world and one external to the discourse event, that is, the discourse-world. Thus, Veronica's repetition of 'you don't get it' may target both Tony Webster in the text-world and the reader in the discourse-world. The fact that, in the review, four out of the six instances of 'get it' report the reviewer's own interpretation of the novel reveals that they recognize the extratextual address implicitly contained in the directly uttered *you*, which transcends ontological domains. In *NLMG*, not only does the phrase 'don't get it' never characterize the homodiegetic narrator, it is also never embedded in a you-narration. Recognizing the extra-textual address in *TSOAE* seems to have led the reviewer to refuse their position as 'unknower'.

The reviewer's disalignment with the homodiegetic narrator can be expanded to include a disalignment with the text-world itself. In fact, on two occasions, the reviewer uses *irrealis* modality to express their desire for a possible world in which the author makes different stylistic choices and in which the fictional characters make different life choices:

> I wish he had not used the words 'history' and 'time' and especially 'memory' in every other sentence – I was suffering from memory lapses myself in the end. [...] There wouldn't, of course, be any Sense of an Ending, or even a beginning, if Veronica had just said, 40 years before (or indeed at any point), 'Look Tony love, this is how it is'.

Because the rest of the review does not feature *irrealis* moods, the use of the conditional in the example quoted above marks, in TWT terms, a world-switch to a 'boulomaic modal-world' (Gavins 2007: 93). In this possible wish-world, the narrator would not use the words 'time', 'history' and 'memory' as much. The intra-action between the narrator and the reader has led the reader to experience the narrator's unreliability, here described in terms of 'memory lapses'. While R1 embraces the version of events given by the homodiegetic narrator and seems to be moulded in the narrator's shape as a result, R2 shows signs of resistance by opening a boulomaic modal-world world through the desiderative verb 'I wish'. R2 admits having been affected by the narrator's words and the text-world despite their will. This statement echoes the bidirectional conceptual metaphor READING IS CONTROL, which leads reader-reviewers to position themselves either as controlling the text or as being controlled by the text (Stockwell 2009: 80). Here, the reviewer casts themself in a passive role, which contrasts with the agentive role construed earlier.

The second possible world, triggered by the conditional clause 'if Veronica had just said', further suggests that the reviewer responds to the novel through Tony's perspective. When the reader's representations of the text-world clash with their preferences, they are likely to create an alternative text-world matching their expectations (Browse 2018: 143). The reviewer takes their distance from Veronica by presenting an alternative world in which Veronica behaves differently. While the reviewer's possible world could be perceived as dissatisfaction with the narrative (hence the two-star review), the reviewer's four-star review of *HF* displays a similar pattern:

> I liked the little insight into what could have been, had Olivia continued to live life with the fast set. Or if Harry had not returned home. Or if several other characters had made countless different choices.So far, so enjoyable, but there are flaws, and one of those is the rather annoying glibness that pervades the novel. Much of this is caused by the over-use of the word 'love' (every other sentence, it seems) so that it starts to feel just a little tacky. The characters keep saying they don't like to use cliches, but then do.

In this passage, the reviewer deplores the prevalence of the word 'love' through the attitudinal lexis 'over-use' and imagines a possible text-world in which Olivia, Harry and other characters make different choices. The alternative fictional universe presents itself as a possible world triggered by the conditional structure if/had + someone + past participle. These patterns seem to resonate with those identified in the review of Barnes's novel and would thus suggest that possible worlds pertain to the reviewer's intra-action with genre conventions. In their review of *TSOAE*, the reviewer creates one boulomaic modal-world and one possible world involving a change of action from one character, Veronica. However, the type of possible world very much depends on the reviewer's reading experience. *HF* features a third-person omniscient narrator; as such, the events presented to us as readers are coloured through several consciousnesses. The novel sketches out a possible world which the reviewer expands on in their review.

In the review of Barnes's *TSOAE*, the possible-worlds (boulomaic modal-world and possible world) emanate from the reviewer in the discourse-world; in the review of *HF*, the possible world occurs across ontological domains at it emerges in the text-world and continues in the discourse-world. The other significant difference has to do with the reviewer's comment on the usage of the words 'time', 'history', 'memory' in *TSOAE* and 'love' in *HF*. The reviewer uses boulomaic modality and associates the modal world with the narrator in their review of *TSOAE* and they use positive polarity embedded in a passive construction in their review of *HF*. Together with the content of each sentence, these features suggest that the use of the word 'memory' (and, to a lesser extent, 'time' and 'history') affected the reader against their will while the use of the word 'love' did not draw the reader in and led them to dismiss this feature as 'cliché'.

In their review of *HF* and *NLMG*, the reviewer directly questions the actions and motives of fictional characters using questions starting with 'why' (in three instances in each review): 'Why does Julia feel she has to give up half her fortune to a man she could conceivably accuse of manslaughter?' (*HF*); 'Why run to this cold, loveless world when they had made their own world, with all its friendships and certainties, from scratch?' (*NLMG*). While these interrogations about the text-world reoccur in the reviews, they cannot be found in the review of *TSOAE*. The repetition of 'why questions' in the reviews led me to question their noticeable absence in *TSOAE*. As far as the review of *TSOAE* is concerned, the reviewer, far from questioning the actions or motives of the narrator, justifies them:

> This was hardly Tony's fault; he was in America at the time. If no-one told him why, how was he supposed to know? And yet we are supposed to feel some sort of sympathy for those who chose not to tell him – just to keep saying, 'you don't get it, do you?' Tony might leave a lot to be desired, but it wasn't his fault he didn't get it.

Here, the reviewer presents mitigating circumstances to explain Tony's behaviour. Interestingly, my stylistic analysis of Barnes's novel reveals that the most salient conceptual metaphor underlying the text is LIFE IS A TRIAL (Doche 2020: 16). Throughout the novel, Webster 'presents himself as a defendant who needs to explain and justify his actions "in front of a court of inquiry"' (Doche 2020: 16), hence the lexical prominence of legalese (see 'corroboration', 'evidence', 'court of law' among other terms). Tony's position as a defendant encourages readers to fulfil the role of juror or judge. Because readers share the thoughts of the defendant, their judgement will depend on whether they accept the narrator's version of events. In the present case, the reviewer refuses to condemn Tony for 'not getting it'. The reader's intra-action with Barnes's novel leads them to accept the position of juror rather than play the role of police investigator by questioning motives.

A diffractive analysis of R2's review shows the places in which their intra-action with *TSOAE* has left its marks. The stylistic features responding to

This study aimed to show how approaching reader-response diffractively could help to account for confounding variables, such as the reviewers' idiosyncratic reviewing practices and the researcher's relationship with the data. Since the study of naturally occurring reader-response does not lend itself to ethnographical considerations, researchers only have textual intra-actions at their disposal, which limits the scope of the inquiry. Moreover, identifying cause and effect ('diffraction patterns') is a time-consuming enterprise. While I have (re)read and analyzed *TSOAE* and read *TF*, *NLMG*, *HF* and *TDG*, neutralizing the reviewers' entanglement with the reviewing genre with complete accuracy would have required me to analyze each of the four novels. One way forward could be to use corpus tools to identify patterns in reviews produced by one reviewer.

By considering other variables/intra-actions, diffractive stylistics has the potential to reduce a major weakness in current reader-response research: the assumption of a one-to-one relationship between reader and text. Diffraction encourages objectivity by recognizing and neutralizing – to some extent – the multiple relationships imbricated in the research. Diffractive stylistics marks a first step from correlation to causation: stylistic variations in readers' reviews allow us to pinpoint with improved precision where the effects of style are being felt.

References

Baker, M. (2021), 'Beyond binaries and before becoming: Reconsidering resistance in UK higher education', *PRISM: Casting New Light on Learning, Theory and Practice*. Available online: https://openjournals.ljmu.ac.uk/index.php/prism/article/view/364 (accessed 17 November 2022).

Bakhtin, M. M. ([1981] 1986), *The Dialogic Imagination: Four Essays*, trans. M. Holquist and C. Emerson, Austin: University of Texas Press.

Barad, K. M. (2007), *Meeting the Universe Halfway: Quantum Physics and the Entanglement of Matter and Meaning*, Durham, NC: Duke University Press.

Barnes, J. (2012), *The Sense of an Ending*, London: Vintage.

Browse, S. (2018), 'From functional to cognitive grammar in stylistic analysis of golding's the inheritors', *Journal of Literary Semantics*, 47 (2): 121–46.

Deleuze, G. and F. Guattari ([1980] 1987), *A Thousand Plateaus: Capitalism and Schizophrenia*, trans. B. Massumi, Minneapolis: University of Minnesota Press.

Descartes, R. ([1641] 1993), *Discourse on Method and Meditations on First Philosophy*, trans. D. A. Cress, Indiana: Hackett Publishing.

Barnes's novel are the actualization – through a readerly perspecti
conceptual metaphor LIFE IS A TRIAL, their position as a knower 'wh
way of focal resonance and the creation of one boulomaic modal-wc
possible world. Although the latter appears to result from the revi
action with the reviewing genre, the alternative world sketched ou
review rests on one character (Veronica) behaving differently. In th
HF, the multiple possible worlds involve various characters makir
decisions. The reviewer's intra-action with the book reviewing g
lead them to question the motives of fictional characters. In that
review of Barnes's marks a difference from repetition since no suc
be observed.

Like R1, the reviewer seems to have experienced the novel t
eyes. R2's engagement can be termed 'dialogic' insofar as their v
expands on and enters in conflict with the voices deployed in
ontological distance between the reviewer and the text – em
reviewer positioning themselves with or against the text-world
frontiers of understanding. R2's review connects fictional languag
The compositions of order are transformed into compositions of
wor(l)d is created (Deleuze and Guattari [1980] 1987: 110). Her
interference from *TSOAE* has led them to go from police investi
their intra-actions with the genre and to accept, resist and open
literary phenomenon without re-enacting it.

Conclusion

Using a diffraction apparatus proved helpful in delaminating wl
intra-actions with *TSOAE* start and when the reviewers' intr
genre stop. R1's 'presumed intimacy' with spatio-temporally d
readers does not pertain to their intra-actions with Barnes's n
reviewer's absorption of Tony Webster's voice and their re-ena
world in the discourse-world emanate from their intra-actic
R2's creation of possible worlds – regardless of star rating –
their intra-actions with the novel. Their intra-actions with *T*
their agentive position as knower, their construal of bouloma
their actualization of the conceptual metaphor LIFE IS A TRI
world. These findings confirm Barad's claim that agency is
intra-actions enact a differential sense of being (2007: 140).

Doche, A. (2020), *Dialogic Strategies and Outcomes in and Around Julian Barnes's The Sense of an Ending: A Linguistic-Stylistic Analysis*, Lyon: Université Jean Moulin Lyon 3.

Doche, A. (2021), 'Hear, here! conversations, equations, translation: On Jonathan Davidson's *A Commonplace*', *Journal of Languages, Texts and Society*, 5: 225–48.

Doche, A. and A. S. Ross (2022), '"Here is my shameful confession: I don't really 'get' poetry": Discerning reader types in responses to Sylvia Plath's *Ariel* on Goodreads', *Textual Practice*, 1–22. Available online: https://www.tandfonline.com/doi/full/10.1080/0950236X.2022.2082516 (accessed 17 November 2022).

Doheny-Farina, S. (1992), *Rhetoric, Innovation, Technology: Case Studies of Technical Communication in Technology Transfers*, Cambridge: MIT Press.

Du Bois, J. W. (2014), 'Towards a dialogic syntax', *Cognitive Linguistics*, 25 (3): 359–410.

Escott, H. (2021), 'Extra-textuality and affective intensities: Moving out from readers to people, places, and things', in A. Bell, S. Browse, A. Gibbons and D. Peplow (eds), *Style and Reader Response: Minds, Media, Methods*, 197–216. Amsterdam: John Benjamins Publishing Company.

Gavins, J. (2007), *Text World Theory: An Introduction*, Edinburgh: Edinburgh University Press.

Genette, G. ([1972] 1980), *Narrative Discourse: An Essay in Method*, trans. Jane E. Lewin, Ithaca: Cornell University Press.

Haraway, D. J. (2016), *Staying with the Trouble: Making Kin in the Chthulucene*, Durham, DC: Duke University Press.

Heidegger, M. ([1927] 1962), *Being and Time*, trans. J. Macquarrie and E. Robinson, Oxford: Basil Blackwell.

Herman, D. (1994), 'Textual "you" and double deixis in Edna O'Brien's "A pagan place"', *Style*, 28 (3): 378–410.

Hidalgo-Downing, L. (2003), 'Negation as a stylistic feature in Joseph Heller's Catch-22: A corpus study', *Style*, 37 (3): 318–40.

Lahey, E. (2004), 'All the world's a sub-world: Direct speech and sub-world creation in "After" by Norman Craig', *Nottingham Linguistic Circular*, 18: 21–8.

Mason, J. (2019), *Intertextuality in Practice*, Amsterdam: John Benjamins.

Martin, J. R. and P. R. R. White (2005), *The Language of Evaluation: Appraisal in English*, London: Palgrave Macmillan.

O'Halloran, K. (2023), 'Posthumanist stylistics', *Language and Literature*, 32 (1). Available online: https://doi.org/10.1177/09639470221140693 (accessed 24 October 2023).

Peplow, D. and S. Whiteley (2021), 'Interpretation in interaction: On the dialogic nature of response', in A. Bell, S. Browse, A. Gibbons and D. Peplow (eds), *Style and Reader Response: Minds, Media, Methods*, 23–41, Amsterdam: John Benjamins.

R1 (2018), 'To begin, wear your watch backwards', *Amazon*, 30 June. Available online: https://www.amazon.co.uk/Sense-Ending-Julian-Barnes/dp/0099564971 (accessed 21 November 2022).

R2 (2011), 'Didn't completely get it', *Amazon*, 30 November. Available online: https://www.amazon.co.uk/Sense-Ending-Julian-Barnes/dp/0099564971 (accessed 21 November 2022).

Rojek, C. (2016), *Presumed Intimacy: Para-social Relationships in Media, Society and Celebrity Culture*, Cambridge, UK: Polity Press.

Sayal-Bennett, A. (2018), 'Diffractive analysis: Embodied encounters in contemporary artistic video practice', *Tate Chapters* 29. Available online https://www.tate.org.uk/research/publications/tate-chapters/29/diffractive-analysis (accessed 17 November 2022).

Stockwell, P. (2009), *Texture: A Cognitive Aesthetics of Reading*, Edinburgh: Edinburgh University Press.

Stockwell, P. (2021), 'In defence of introspection', in A. Bell, S. Browse, A. Gibbons and D. Peplow (eds), *Style and Reader Response: Minds, Media, Methods*, 165–78, Amsterdam: John Benjamins.

3

Cohesion and Solidarity in Covid-related Addresses to the Nation

Chris Fitzgerald and Helen Kelly-Holmes

Introduction

Covid-19 was not only an unprecedented global public health event, it also resulted in a major global discourse challenge for a number of reasons. Along with medical measures, health communication was seen as 'a key and necessary factor in saving lives' during the crisis in order to help societies to handle the uncertainty of the crisis, manage fear and 'promote and accomplish adherence to necessary behaviour change, and meet individuals' fear and foster hope in the face of a crisis' (Finset et al. 2020: 873). As well as health information, 'prime time speeches to the people by kings, presidents, prime ministers and religious leaders' (Finset et al. 2020: 873) were seen as a central aspect of this unprecedented flow of health information (Finset et al. 2020). Ireland was no exception here and followed the global pattern of regular 'address to the nation' type events at pivotal points in the crisis, daily updates from the health authorities, radio and TV advertising campaigns, and the provision of resources and information for public institutions and general public.

On Saint Patrick's Day (17 March 2020), Leo Varadkar, the Taoiseach (Prime Minister) of Ireland, delivered a speech announcing the details of the first lockdown that Ireland would face of the Covid-19 era. This would be the first of many addresses to the nation aimed at delivering updates, encouragement and justification for Covid regulations to the Irish people. As the impact of the virus increased and waned over the subsequent two years, such addresses became commonplace and an expected feature of public discourse as the Taoiseach changed from Leo Varadkar to Micheál Martin.

Guidance soon began to emerge about how to manage this flow of information and what the features of a successful public discourse around Covid-19 might include. It was widely recognized that Covid-19 'has introduced unique challenges for health communication' (Ratzan, Sommarivac and Rauh 2020: 2) and that a streamlined approach would be needed, particularly in light of the flow of misinformation. The following can sum up the guidance:

1. Stick to the facts – declare what is known and unknown.
2. Be consistent and specific – limit the number of spokespeople in order to 'establish trusted leadership' (Ratzan, Sommariva and Rauh 2020: 1). What is interesting here is that the advice about limiting the number of spokespeople led to criticism in Ireland and probably elsewhere about the lack of diversity in public health emergency team.
3. Demonstrate the ability to cope with and function in uncertainty.
4. Acknowledge emotion, since empathy is needed 'to dispel panic and passivity'. 'Adopt an empathic style of communication to grab public attention and address health concerns. Empathy in communication is critical for managing public anxiety and promoting behavioural complication with public health guidelines' (Mheidly and Fares 2020: 416).

While facts were important, it was also deemed important to promote a discourse of 'optimistic anxiety' aimed at reaching an ideal balance whereby 'citizens must be anxious enough to take the advice from authorities to heart and optimistic enough as to feel that their actions make a difference' (Petersen in Finset et al. 2020: 874).

Efforts to combat the spread of Covid-19 are dependent on cooperation between citizens and government. However, evidence from previous pandemics indicated that 'official recommendations are often met with skepticism by many' (Finset et al. 2020: 874). Therefore, along with empathy and 'optimistic anxiety', the fostering of a collective spirit was identified as being particularly crucial: 'Even if citizens are more isolated than ever, in more or less self-imposed quarantines, appeals to collective action and a spirit of we-are-in-it-together are very important to flatten the curve and reduce the rate of infection' (Finset et al. 2020: 874). This could be achieved by 'appeals to solidarity and shared responsibilities' (Finset et al. 2020: 874) and 'demonstration of concern by role models' (Finset et al. 2020: 874).

Our focus in this chapter is on how these appeals to solidarity were discursively constructed and performed in Address to the Nation type events in the Irish context of Covid-19 public health discourse. To achieve this, we employ the following methods:

1. A corpus-driven approach is used to establish the salient characteristics of the speeches.
2. Content analysis is used to analyze figurative language.
3. An approach derived from Critical Discourse Analysis is utilized to analyze the public and media reaction to the use of one particular solidarity-building device, that is, the use of poetic quotation in the speeches.

The Irish Covid Speeches Corpus

To establish a linguistic characterization of the addresses to the nation, this study takes a corpus-driven approach (Tognini-Bonelli 2001: 87). To achieve this, a corpus of these speeches was constructed from the initial speech on 17 March 2020, through twenty speeches up to 29 April 2021. Transcriptions of these speeches, ranging in length from 629 words to 2,244 words were accessed from the Government of Ireland website (www.gov.ie) and compiled into a corpus, the Irish Covid Speeches Corpus (ICSC). Table 3.1 outlines a list of the speech dates and number of words per speech. The speeches which are shaded are those delivered by Leo Varadkar, while the remainder are those delivered by Micheál Martin.

Table 3.1 Dates and word-count of speeches in the Irish Covid Speeches Corpus

Date	Words
17 March 2020	1829
24 March 2020	1702
27 March 2020	1260
10 April 2020	629
1 May 2020	1494
5 June 2020	1139
19 June 2020	869
15 July 2020	1122
4 August 2020	1091
7 August 2020	1228
18 August 2020	888
18 September 2020	1205
24 September 2020	890
5 October 2020	1032
19 October 2020	1409

Date	Words
27 November 2020	1679
30 December 2020	2244
26 January 2021	827
23 February 2021	1537
30 March 2021	1224
29 April 2021	1672
Total	26970

The ICSC was analysed via the corpus software package AntConc (Antony 2022) to determine salient items and to provide an objective overall linguistic characterization of the corpus based on quantitative results using keyword lists and lists of frequent single and multi-word items. Keyword searches (Scott 1997) offer the researcher insights into what is unique about a corpus when studied in relation to a reference corpus. According to Culpeper and Demmen (2015: 90), 'in corpus linguistics, a keyword has a quantitative basis: it is a term for a word that is statistically characteristic of a text or set of texts.' Scott and Tribble (2006) look at two aspects of keywords which emerge in this study, those keywords that relate to 'aboutness', that is, keywords which one may be likely to predict and keywords which relate to style, those that are key due to the particular genre of the language under analysis. In addition multi-word units are analyzed as they are described as a means of revealing patterns of language to observe variety across registers and discourse types (see Cortes 2002; 2004; Biber et al. 2004; Hyland 2008; Csomay 2012). Further to this, concordances (Sinclair 1991) are used to analyse items in further context beyond single- and multi-word units.

From the quantitative analysis using corpus linguistics tools, this study shifts to a more qualitative approach based on content analysis to determine and analyse figurative expressions. Content analysis, according to Ary et al. (2018) is a research method applied to written or visual materials for the purpose of identifying specified characteristics of the material. The system for characterizing and analysing figurative expressions used in this study largely follows that of Nainggolan et al.'s (2021) procedure for the analysis of figurative language in Joe Biden's American Presidential victory speech:

1. Close reading of speeches;
2. Identify instances of figurative language;
3. Classify figurative language into types of figurative expression;
4. Classify figurative expressions into themes for further descriptive analysis.

Step 3 was carried out using Crystal's (1991: 166) broad definition of figurative language, in which he describes it as 'an expressive use of language where words are used in a non-literal way to suggest illuminating comparisons and resemblances'. The classification of language as literal or figurative has proven challenging with different approaches taken to this classification based on various definitions of literality. We take a broad perspective, classifying items as figurative if they clearly fall within the categories described by Leech in Dewi (2010: 2). Leech identifies eight categories of figurative language: personification, simile, metaphor, hyperbole, irony, litotes, metonymy and oxymoron. Of these, irony, litotes, metonymy and oxymoron are absent in the data. Thus, the four types categorized are personification, simile, metaphor and hyperbole.

Hyperbolic expressions proved challenging to determine due to the context in which the speeches were delivered and so a further model was employed. Burgers et al. (2016: 166) define hyperbole as 'An expression that is more extreme than justified given its ontological referent'. They describe one of the main determinants of hyperbole as being exaggeration and provide a model for the identification of hyperbole which was adhered to in this study. Statements that were deemed factual, for example, the Covid-19 era being described as 'unprecedented', were not characterized as hyperbolic, but those which either have no ontological referent or are exaggerated beyond literal interpretation were deemed hyperbolic, such as (from 19 June 2020 speech), 'Some have asked whether there is a limit to what we can achieve. My answer is that the limit does not exist.' While some studies, such as Kunneman et al. (2015), include intensifiers such as adjectives like *fantastic* in their characterization of hyperbole, these were not included here as these are not seen to be exaggerations in this context and not aligning with the above-cited definition of figurative language.

In addition to these categories of figurative language, we analyse poetic quotation as a strategy aimed at promoting solidarity with Irish people. We focus on the use of quotation and give particular attention to a quotation of a Seamus Heaney poem in Leo Varadkar's speech of 10 April 2020. For this, we expand our analysis beyond the text and look at the responses to this in social media (Twitter) and traditional media (the *Irish Times*). To achieve this, we employ approaches derived from Fairclough's (1989; 1995) Critical Discourse Analysis (CDA) which proposes three phases of analysis:

1. Description using text analysis,
2. Interpretation via processing analysis,
3. Explanation via social analysis.

Salient characteristics of the ICSC

A reading of the keywords of the ICSC with the British National Corpus (BNC) as a reference corpus in Table 3.2 shows that the ten most frequent keywords show a salience of Covid-related items (in bold). However, more noteworthy is the prominence of Irish language items (in italics), which are common features of Irish political speeches, usually employed at the opening and closing of speeches. It is perhaps unsurprising that the acronym for the National Public Health Emergency Team is ranked first as this is particularly related to the 'aboutness' (Scott and Tribble 2006) of this discourse. These quantitative results show the attempt to achieve the balance identified above between information and facts on the one hand (the prevalence of Covid vocabulary) and solidarity-building on the other (the prevalence of Irish language items). The Covid vocabulary identified is new, specific and common to the emerging global Covid lexicon (e.g. vaccinate, re-open, non-essential) that has come to characterize warnings, directions and informational discourse. On the other hand, the Irish language items are drawn exclusively from phatic discourse and point to continuity with public speaking and national solidarity building.

Shifting from single-word items to multi-word units reveals more about the prevalent themes of the speeches. The ten most frequent three-word units listed in Table 3.3 reveal the salience of a theme of solidarity and togetherness.

Table 3.2 Top ten keywords in the Irish Covid Speeches Corpus

Rank	Word
1	**NPHET**
2	*Agaibh*
3	**Non-Covid**
4	*Maith*
5	*Raibh*
6	**Reopen**
7	**Non-essential**
8	*Agus*
9	**Vaccinate**
10	**Re-open**

Table 3.3 Top ten three-word units in the Irish Covid Speeches Corpus

Rank	Three-word units
1	of the virus
2	we need to
3	I want to
4	the spread of
5	be able to
6	the end of
7	of the disease
8	that we have
9	the number of
10	one of us

While many of these items are related to the spread of the virus (*of the virus, the spread of, of the disease*) and, upon inspection of further context, reveal an urgency to the curtailment of the virus, there are items here which are worthy of highlighting in further context. Number 10 above, *one of us*, reveals a prominent theme when observed in context. When observed in the further context in concordance lines (see Table 3.5), this is usually used to express a degree of personal and social responsibility and to present a sense of togetherness. This could be perceived as a means of deflecting responsibility from the state by dint of shifting this onto citizens. Pronoun usage in public discourse such as this is highly meaningful (see e.g. Fairclough 1989). As expected for a political address of this kind, the first personal plural pronouns 'we', 'our' and 'us' feature most frequently, with the function of delineating and unifying the nation. Within the Covid-19 context, the national 'we' being delineated here draws a boundary around those citizens who wish to comply with health guidance, who agree with government strategy and who wish to bring an end to the pandemic.

While the first personal plural pronouns also dominated the general health promotion communication ('We're all in this together'), the first-person singular 'I' did not feature, unlike in the ICSC. For this type of address, the speaker's identity, role or office (e.g. Prime Minister, Minister for Health) is of course crucial. They have the authority to use their identified and identifiable voice and in fact this is essential for these texts to have meaning, and thus we find frequent use of first-person singular 'I' to express empathy and understanding, to make commitments and guarantees (e.g. 'Tonight I want you to know'), and to lead by example.

Table 3.4 Frequencies of pronouns used in the Irish Covid Speeches Corpus

	Pronoun	Frequency
1	we	718
2	our	319
3	it	263
4	i	199
5	you	128
6	us	96
7	their	85
8	your	71
9	they	61
10	them	46

Table 3.5 Concordance of *one of us* in the Irish Covid Speeches Corpus

staff … big demands of every single	one of us	Tonight I want you to know
What happens now is up to each	one of us	. Show your support to our
you are afraid of the virus every	one of us	can have an impact on the lives
we have shown that we can beat it. Each	one of us	has the power to suppress it
friends and our communities, every	one of us	has a responsibility to protect
in other areas. Each and every	one of us	needs to reflect on that.
sacrifices that are being asked of each	one of us	that we can slow the new.

Figurative language in the Irish Covid Speeches Corpus

Before looking in more depth at figurative expressions in the corpus, it is of interest to highlight where these expressions occur in the speeches. The geography of the political speech and its generic conventions are well known to the public; however, the address to the nation is less well known, albeit Covid-19 ushered in an unprecedented number of such addresses. It is also likely that while the public can draw on interdiscursive knowledge of a small number of such addresses that have occurred previously, they may also have been exposed to the genre through media discourse, even fictional, and so some degree of interdiscursivity can be relied on. Unlike normal political speeches, the address to the nation Covid-19 speeches took place in formal settings – generally the official government buildings. The addresses were made from a podium which had the national symbol, the harp and both the Irish flag and the EU flag were also part of the setting. In general, the Taoiseach, or prime minister, emerged (presumably from an important meeting) and walked along a carpet to the podium to deliver the address to the waiting media, who were physically preset, and the public who were watching from their homes in isolation. Initially, the Covid addresses were key moments, eagerly anticipated by the public and avidly watched, but as the pandemic went on, they lost some of their ability to bring the nation together.

The structural form of the speeches follows a typical pattern of six stages that is outlined below with examples from the speeches of typical occurrences of these stages:

1. *Greeting*: 'Good evening'
2. *Contextualization of the speech with reference to the previous speech*:
   'Almost six weeks ago, I stood here and talked with you about what urgent action we needed to take together to drive down high and rising Covid infection rates.'
3. *Reference to the positive/negative changes in the status of the Irish situation regarding Covid*:

   There is a day-on-day increase in the number of admissions to intensive care units, and the number has doubled since Monday.
4. *Outline of how the state will address the positive/negative changes set out in 3*:

   So, with effect from midnight tonight, for a two-week period until Easter Sunday, 12 April, everybody must stay at home in all circumstances.
5. *Comment on citizens' efforts and words of encouragement*:
   Working together our country will come through this Emergency. We will be tested – but will succeed.
6. *Closing*: 'Thank you'

As we might expect in political discourse of this nature, figurative language features highly. The pandemic was a unique event and these speeches were designed to give certainty and reassurance about something which very few understood. Much of the figurative language and phatic Irish occurs in stages two and five, that is, just after the opening of the speech and just before the closing.

As has been stated, of Leech's (1979 in Dewi 2010: 16) categories of figurative language, four types were found in the data, with examples below:

> Metaphor: *This is the calm before the storm – before the surge.*
> Simile: *She said her patients were like family; she said 'they call us their best friends'.*
> Hyperbole: *Thanks also to everyone helping others in a million different ways.*
> Personification: *It is constantly on the search for new people to infect.*

Overall, 111 figurative expressions were identified. As seen in Table 3.5, metaphor is the most common type of figurative expression in the data, which tallies with the findings of Nainggolan *et al.* (2021). Metaphors are a key feature of political discourse, and given the nature of the pandemic, metaphors clearly had an important role to play in making the complex situation comprehensible for the public. The type and role of metaphorical expression emerging during the pandemic in public and media discourse have been identified as a particularly salient feature of pandemic discourse (Charteris-Black 2021; Semino 2021; Hanne 2022).

The twelve instances of personification found relate to the virus. The verbs and adjectives collocating with the virus give it qualities that are mostly used to describe actions and characteristics of people. The virus is described as being *cruel, unrelenting* and *indiscriminate*, it *doesn't care who you are, or where you're from. It just wants to spread.* The virus is said to *attack by stealth* and to be *constantly on the search for new people.* This personification constructs the virus as an enemy that has the capacity to attack and one which must be attacked in return. It provides substance to a narrative of the state and individuals being in a battle with a common enemy that can be defeated if people adhere to the rules.

Table 3.6 Occurrences of figurative expressions in the Irish Covid Speeches Corpus

Types of figurative expression	Occurrences
Metaphor	87
Personification	12
Simile	6
Hyperbole	6
Total	111

This theme of a fight or battle comes across strongly in much of the metaphorical language used in the speeches.

This personification of the virus as an attacking enemy or even an alien is reinforced by metaphors of battle/fighting used in the speeches. There are three instances of having *our guard up* mentioned as well as the word *fight* occurring three times in figurative expressions. In addition, such expressions as *we can beat it* and *testing remains the key weapon in our armoury* are used to invoke a sense of a battle between the people and the virus. The need for a common 'other' is key to constructing the nation ('us') in contrast to an enemy. Such figurative language utilizing the lexical field of war/battle reinforces the direct and indirect imperatives in the speeches and in the general public health discourse. This type of metaphor has also been found in other contexts, as can be seen in Rajandran's (2020) analysis of metaphors for Covid-19 used by Singaporean and Malaysian leaders.

The most common metaphorical expression in the data is associated with a type of road or path, again unsurprising in political discourse of this kind. There are fourteen metaphors invoking a path or road in the data. These include

1. we have a long road ahead of us,
2. there will be bumps in the road,
3. the road ahead will continue to have many turns.
4. the road we are on is hard. But it is the road we must take, together,
5. the vaccination programme will lay the path out of this pandemic,
6. we must apply the brakes to movement,
7. we are on the final stretch of this terrible journey.

This theme is one which the Irish government integrated into the title of its plan for managing the pandemic, 'The Path Ahead'. The use of these expressions makes the process comprehensible and provides a sense of a gradual process, which is aided by the various influences on a way out: the compliance of citizens, the roll-out of vaccinations and state-run procedures such as testing and tracing. These expressions are used to manage expectations and to encourage patience as the pandemic continues to disrupt daily lives. Other metaphors used invoke darkness and light (*better and brighter days that lie ahead*) and storms (*a calm before the storm*) exemplify the need for 'optimistic anxiety' in Covid-19 discourse, discussed above.

Another device used to build solidarity was the Irish language. While Irish is the first official language in the Constitution and in accordance with related language rights legislation, most notably the Official Languages Act 2003, public health information must be made available in Irish. Therefore, public health notices and television and radio advertising provided parallel information and advice in Irish. However, in the solidarity-building address to the nation type speeches, Irish was largely confined to figurative or symbolic purposes. For example, words of welcome ('phatic Irish'), which is expected in these types of genres in Ireland, and also the use of sayings in Irish or rallying words at the end of the speeches.

This symbolic usage of Irish is not trivial. Coupland, Bishop and Evans (2006) use the term 'ceremonial language' to describe the use of minoritized languages to delineate both important national public events (sporting events, key moments such as elections, and to mark dates in the political and national year) and/or significant personal events (birthday greetings, condolences). Even where the language is not in everyday usage by the speaker and/or the majority of addressees, it is still highly meaningful and not to be dismissed. Ceremonial Irish was therefore an important part of the Covid-19 addresses, even where it only counted for a small part of each individual speech. Another important communicative device was the intertextuality created by the quoting of poetry, prose and ancient Irish proverbs, and we turn to a discussion of the reception of this strategy now.

Public response to the use of figurative language

While figurative and poetic language was tolerated and even celebrated in the early stages of the pandemic when the governing party's approval ratings were correspondingly high, this tolerance abated with the public's overall Covid-related fatigue. This turning point is marked following the 10 April 2020 speech by the Taoiseach Leo Varadkar, who referenced the Nobel laureate Irish poet Seamus Heaney in both the opening and closing phases of his speech. In the opening of the speech, Mr. Varadkar referred to the troubles in Northern Ireland, stating that:

> During the worst year of those Troubles the poet Seamus Heaney spoke about what was happening and predicted that "if we winter this one out, we can summer anywhere". I know these words have provided inspiration to many Irish people as we deal with this Emergency. They remind us that we are in this together, we can get through it, and better days will come.

Mr. Varadkar returned to Heaney in the closing phase of the speech to invoke togetherness:

> In one of his best collections of poems, Heaney celebrated the human chain of help that can bring about an almost miraculous recovery. As Heaney wrote, we were "all the more together for having had to turn and walk away". In the days ahead we must continue to turn and walk away from each other and from doing the things we would like to do. But we will be all the more together for having done so.

Both traditional and social media reacted to this quotations. An article in the national paper of record, the *Irish Times*, by leading political commentator Miriam Lord, known for her witty takedowns and wry observations (https://www.irishtimes.com/news/politics/miriam-lord-messages-of-hope-but-no-bank-holiday-weekend-rollover-jackpot-1.4226393), highlights the backlash against Mr. Varadkar's use of Heaney. The terminology and tropes of the pandemic which the public have now become familiar with, are used to construct an ironic and humorous critique of the use of poetry by the Taoiseach. In the article, Lord refers to a 'poetry pandemic', states that 'the Taoiseach has succumbed completely to the Heaney bug and is now a super-spreader' and suggests that the Taoiseach 'must poetically distance himself immediately. Because this thing is infectious.' The article finishes with Lord paraphrasing Heaney's (1966) *Digging*: 'Between my finger and my thumb The squat pen rests I'll dig lumps out of Leo with it Yet.' The criticism of the repetitive use of such poetry captures the parallel sense of Covid-related fatigue that was widespread at the time of its writing as lockdowns were being extended and preventative measures were becoming overly familiar.

In addition to reaction in traditional media, social media commentary also critiqued Varadkar's use of poetic quotation. Twitter content often expresses extreme opinions and is seen as a gauge of polar perspectives (Yaqub et al. 2018) but does present a rich source for investigation into public sentiment towards topical issues (Sardinha 2022) and emerging public reaction to a variety of topics (Bian et al. 2016). A review of tweets referring to Mr. Varadkar's use of Heaney reflects the polarity of the platform and the often-extreme language that is one of the hallmarks of Twitter discourse. Of the twelve tweets identified that were posted after the delivery of Mr. Varadkar's speech that refer directly to the use of poetry, three were positive in their assessment of his use of the poem and nine were negative. Contextualizing the tweets within broader societal issues using Fairclough's (1989) model reveals that many of the tweets use reference to the use of poetry to frame other ideological points of argument.

An example of a positive tweet reflects the presumed rationale for the use of poetry by Mr. Varadkar: 'I'm glad we have a Taoiseach who quotes poetry when updating the country on #COVID19 Good choice of Heaney today. I wonder has Dominic Raab ever quoted poetry?'

The negative view of the contemporary British spokesperson, Dominic Raab, is an important indication of support for the choice of poem used by Mr. Varadkar – emphasizing the superiority of 'us' (Irish people who appreciate and respond to poetry) over 'them' (the British public). The tweets that express a negative sentiment towards the use of poetic quotation highlight a contrast between imposing lockdowns while quoting poetry: 'Lockdown until 5 May and Leo's banging on about Seamus Heaney's poems fuck off im sick of this poxy country'.

Three of the nine negative tweets directly refer to a lack of funding for the creative arts, while utilizing a creative medium (poetry) in a political context: 'Interesting to note @LeoVaradkar quoting Seamus Heaney in his address to the nation. Soon politicians won't have our beautiful, INTERNATIONALLY RENOWNED POETS AND LITERATURE to use in political speeches if they don't FUND THE ARTS'.

In these tweets, the argument against the use of poetry in this context is used to highlight a wider societal issue that the authors have issue with (lockdowns or funding for the arts). These tweets represent what Goffman (1974) calls 'integrated frames' as they utilize one topic to frame a criticism of another somewhat related topic. A criticism of the use of poetry in the speech provides the opportunity to frame a broader ideological stance, for example, funding for the arts.

Discussion

The Covid-19 pandemic was not just a significant and unprecedented event for health globally, its impact resonated widely across all sectors of society. Communication was a key 'weapon' to use the preferred metaphors of the discourse, in the arsenal of governments attempt to 'battle' the pandemic. As the current study and related studies have shown, the pandemic has created a new lexicon and discourse, which has both globally common elements as well as highly localized ones. The Irish Covid Speeches Corpus exemplifies the attempt by governments globally to follow the directions of balancing information with empathy and offering hope while maintaining a necessary level of anxiety. As we have seen, pronoun choice to enhance and delineate the nation of good, compliant citizens, figurative language particularly utilizing the lexicon of war, phatic Irish and intertexual linkages to a national canon of poetry, prose and proverbs constitute key elements in this discourse in the Irish context.

The potential ambiguity that the use of figurative language might cause runs counter to the recommendations for communicating during a pandemic as outlined earlier in the chapter. However, we have seen that frequent use of figurative language in the speeches analysed constructs a narrative of a battle against a virus that is akin to a battle of good versus evil and that figurative expressions associated with a journey or road help make the situation comprehensible, communicate a plan and decisiveness in the context of uncertainty, and create an impression of progress that is reliant on the compliance of citizens. Such usage also needs to be balanced, as per public health communication guidelines discussed above, with unambiguous facts and information.

Negative public and media reaction to the use of poetry in the speeches emerged when certain poets were overused and when attention was not being paid to the timing of the use of poetry. Reaction shows that, rather than creating a sense of unity, at a time of rising case numbers and increasing hardship for workers in certain sectors, the use of poetry represents a disconnect with the pain felt by those suffering the consequences of a pandemic.

While the creative realm of poetry may not be deemed a natural synthesis with the rigid structures of political speech-making, Orr (2008: 412) contends that politics is 'the most favourable non-artistic arena for a certain type of poetic sensibility'. Flint (1996: xii) argues that the two are 'inescapably linked', stating that 'Theoretical reflection about the relationship between poetry and politics is a necessary activity: one that forces us to interrogate the nature of the coupling, itself rhetorically enacting, through the ready neatness of its alliteration, the apparent indissolubility of the bond.' This bond is one which should be carefully sealed and consideration given to the timing and overuse of poetic quotation in political rhetoric.

Reflection on the views of poets towards the use of their poetry in political contexts might be a consideration for politicians when choosing to use poetry. In 'The Anxiety of Influence', O'Brien (2005) describes the impact on both W.B. Yeats and Seamus Heaney of the prospect of their poetry having an influence on political matters. Despite their reticence towards their poetry permeating political rhetoric, their public popularity and prolific political thematic inclinations perhaps made it inevitable that their words would echo on the plinths of politicians with a penchant for poetry.

Finally, the response to the use of figurative language and poetic quotation in the pandemic seems to indicate that an ongoing awareness of the real-time societal conditions of reception (see Fairclough 1989) is crucial (timing), as well as a constant monitoring of the manifest and constitutive intertextuality (see Fairclough 1989) within the textual chain of these speeches to ensure there is sufficient but not excessive use and repetition.

References

Amador-Moreno, C. P. (2010), 'How can corpora be used to explore literary speech representation', in A. O'Keeffe and M. McCarthy (eds), *The Routledge Handbook of Corpus Linguistics*, 531–44, Abingdon/New York: Routledge.

Anthony, L. (2022), AntConc (Version 4.1.4) [Computer Software]. Tokyo, Japan: Waseda University. Available online: https://www.laurenceanthony.net (accessed 14 February 2023).

Ary, D., L. C. Jacobs, C. K. S. Irvine and D. Walker (2018), *Introduction to Research in Education*, Boston: Cengage Learning.

Bian, J., K. Yoshigoe, A. Hicks, J. Yuan, H. Zhe, X. Menjun, Y. Guo, M. Prospen, R. Salloum and F. Modavem (2016), 'Mining twitter to assess the public perception of the "Internet of things"', *PLoS ONE*, 11 (7): 1–14.

Biber, D., S. Conrad and V. Cortes (2004), 'If you look at … : Lexical bundles in university teaching and textbooks', *Applied linguistics*, 25 (3): 371–405.

Burgers, C., B. C. Brugman, K. Y. Lavalette and G. J. Steen (2016), 'HIP: A method for linguistic hyperbole identification in discourse', *Metaphor and Symbol*, 31 (3): 163–78.

Charteris-Black, J. (2021), *Metaphors of Coronavirus: Invisible Enemy or Zombie Apocalypse?*, Cham: Palgrave Macmillan.

Cortes, V. (2002), 'Lexical bundles in Freshman composition', in R. Reppen, S. M. Fitzmaurice and D. Biber (eds), *Using Corpora to Explore Linguistic Variation*, 131–45, Masterdam: John Benjamins.

Cortes, V. (2004), 'Lexical bundles in published and student disciplinary writing: Examples from history and biology', *English for Specific Purposes*, 23 (4): 397–423.

Coupland, N., H. Bishop, B. Evans and P. Garrett (2006), 'Imagining Wales and the Welsh language: Ethnolinguistic subjectivities and demographic flow', *Journal of Language and Social Psychology*, 25 (4): 351–76.

Crystal, D. (1991), *A Dictionary of Linguistics and Phonetic*, Cambridge: Basil Blackwell Ltd.

Csomay, E. (2012), 'Lexical bundles in discourse structure: A corpus-based study of classroom discourse', *Applied linguistics*, 34 (3): 369–88.

Culpeper, J. and J. Demmen (2015), 'Keywords', in D. Biber and R. Reppen (eds), *The Cambridge Handbook of English Corpus Linguistics*, Cambridge: Cambridge University Press, 90–105.

Dewi, Kumala Sari (2010), *An Analysis of Figurative Meaning in The Time's Magazine's Advertisement*, Medan: Universitas Sumatra Utara.

Fairclough, N. (1989), *Language and Power*, London: Longman.

Fairclough, N. (1995), *Critical Discourse Analysis*, London: Longman.

Finset, A., H. Bosworth, P. Butow, P. Gulbrandsen, R. L. Hulsman, A. H. Pieterse, R. Street, R. Tschoetschel and J. van Weert (2020), 'Effective health communication – a key factor in fighting the COVID-19 pandemic', *Patient Education and Counseling*, 103 (5): 873–6. Available online: https://doi.org/10.1016/j.pec.2020.03.027 (accessed 14 February 2023).

Flint, K. (1996), *Poetry and Politics*, Suffolk: D.S. Brewer.

Gibbs Jr, R. W., D. L. Buchalter, J. F. Moise and W. T. Farrar IV (1993), 'Literal meaning and figurative language', *Discourse Processes*, 16 (4): 387–403.

Goffman, E. (1974), *Frame Analysis: An Essay on the Organization of Experience*, Boston: Harvard University Press.

Hanne, M. (2022), 'How we escape capture by the "war" metaphor for COVID-19', *Metaphor and Symbol*, 37 (2): 88–100.

Hyland, K. (2008), 'As can be seen: Lexical bundles and disciplinary variation', *English for Specific Purposes*, 27 (1): 4–21. https://www.irishtimes.com/news/politics/miriam-lord-messages-of-hope-but-no-bank-holiday-weekend-rollover-jackpot-1.4226393 (accessed 14 February 2023).

Kunneman, F., C. Liebrecht, M. van Mulken and A. van den Bosch (2015), 'Signaling sarcasm: From hyperbole to hashtag', *Information Processing & Management*, 51 (4): 500–9.

Leech, G. (1979), *A Linguistic Guide to English Poetry*, New York: Longman.

Mheidly, N. and J. Fares (2020), 'Leveraging media and health communication strategies to overcome the COVID-19 infodemic', *Journal of Public Health Policy*, 41 (4): 410–20. https://doi.org/10.1057/s41271-020-00247-w

Nainggolan, F., D. A. Siahaan, B. Sinurat and H. Herman (2021), 'An analysis of figurative language on Joe Biden's victory speech', *International Journal on Integrated Education*, 4 (3): 364–75.

O'Brien, E. (2005), 'The anxiety of influence: Heaney and Yeats and the place of writing', *Nordic Irish Studies*, 4: 119–36.

Orr, D. (2008), 'The politics of poetry', *Poetry*, 192 (4): 409–18.

Rajandran, K. (2020), '"A Long Battle Ahead": Malaysian and Singaporean prime ministers employ war metaphors for COVID-19', *GEMA Online Journal of Language Studies*, 20 (3): 261–7.

Ratzan, S. C., S. Sommariva and L. Rauh (2020), 'Enhancing global health communication during a crisis: Lessons from the COVID-19 pandemic', *Public

Health Research & Practice, 30 (2). Available online: https://doi.org/10.17061/phrp3022010 (accessed 14 February 2023).

Sardinha, T. B. (2022), 'Corpus linguistics and the study of social media: A case study using multi-dimensional analysis', in A. O'Keeffe and M. McCarthy (eds), *The Routledge Handbook of Corpus Linguistics*, 656–74, London: Routledge.

Scott, M. (1997), 'PC analysis of key words—and key key words', *System*, 25 (2): 233–245.

Scott, M. and C. Tribble (2006), *Textual Patterns: Key Words and Corpus Analysis in Language Education*, Amsterdam: John Benjamins Publishing.

Semino, E. (2021), '"Not soldiers but fire-fighters" metaphors and covid-19', *Health communication*, 36 (1): 50–8.

Sinclair, J. (1991), *Corpus, Concordance, Collocation*, Oxford: Oxford University Press.

Tognini-Bonelli, E. (2001), *Corpus Linguistics at Work*, Amsterdam: John Benjamins.

Yaqub, U., N. Sharma, R. Pabreja, S. A. Chun, V. Atluri and J. Vaidya (2018), 'Analysis and visualisation of subjectivity and polarity of Twitter location data', in A. Zuiderwijk and C. C. Hinnant (eds), *Proceedings of the 19th Annual International Conference on Digital Government Research: Governance in the Data Age*, 1–10, New York: Association for Computing Machinery.

4

Using Think-Aloud Data to Explore Pathetic Fallacy's Impact on Narrative Empathy

Kimberley Pager-McClymont and Fransina Stradling

Introduction

This chapter illustrates the benefit of using reader data to examine the experiential impact of conceptual metaphor mappings on empathy. To do so, we reflect on the methodological approach we used in a previous study (Stradling and Pager-McClymont 2023) to consider pathetic fallacy (PF), a specific type of conceptual metaphor, as a potential trigger of narrative empathy in Walker's *The Flowers* (1973). In this study, we employed a think-aloud protocol to analyse participants' perception of PF in dedicated paragraphs, and the impact this had on participants' empathetic engagement. This methodological approach is unique in three ways:

1. No previous research has used the same dataset for analysing both empathy and perception of a textual trigger (here PF);
2. No protocol for analysing readers' perception of PF exists, so we created our own;
3. We were the first to apply Fernandez-Quintanilla's framework (2018) for identifying evidence of empathy in think-aloud data.

While an overview and data for this study can be found in greater detail in Stradling and Pager-McClymont (2023), this chapter argues that our methodology provides an effective approach to studying the role of textual triggers on empathy. By situating our methodology in the existing landscape of stylistic approaches to narrative empathy, we seek to answer the following research questions (RQs):

- RQ1: How might one empirically examine the experiential impact of conceptual metaphor mappings on readers?
- RQ2: What are the affordances of think-aloud data for studying the role of a particular textual feature in shaping narrative empathy?

We first provide a literature review of existing research on linguistic features' impact on narrative empathy, as well as a review of reader response data and think-aloud protocol. We then illustrate how the combination of our think-aloud data and our analytical methods to analyse PF and readers' perception of it as well as readers' empathetic engagement in paragraphs of the story featuring PF contributed to our conclusions. Lastly, we reflect on our arguably unique methodology and offer insight as to what impact this could have on future research.

The stylistic exploration of narrative empathy

We define narrative empathy as 'an online[1] cognitive and/or affective response to narrative characters that is similar in nature to one's perception and understanding of the stimulus emotion' (Stradling and Pager-McClymont 2023).[2] Furthermore, we consider empathetic engagement to be shaped by textual and readerly features, including readers' experiential background and affective dispositions. Arguably, research into narrative empathy thus calls for analysis of linguistic features in conjunction with a detailed exploration of reader response data, which (cognitive) stylistics is ideally suited for (see Whiteley and Canning 2017 for an overview of reader response research for stylistic purposes).

However, stylistic exploration of narrative empathy is still emergent. The first to posit a list of narrative techniques and their textual realizations that may contribute to narrative empathy is Keen (2006). Based upon theoretical explorations of narrative empathy, this list includes characterization techniques, descriptions of narrative situation, plot devices, use of generic and formal choices that may be foregrounded to the reader and paratextual features such as texts' length and genre expectations. Little empirical research existed then to verify the contribution of these textual features to narrative empathy, and little has been published since. Exceptions include László and Smogyvári (2008), Van Lissa et al. (2016), Kuzmičová et al. (2017) and Fernandez-Quintanilla (2020).[3] For instance, László and Smogyvári (2008) test whether there would be a difference in narrative empathy if a story character was part of a readers' in-group or out-group. Using Likert scales and a narrative recall task, they find no significant differences between these two conditions.

Van Lissa et al. (2016) study the impact of narrative perspective on narrative empathy and trust by presenting two different groups of participants with the same short story either in its original first-person perspective or a manipulated version in third person. They do not find a difference in empathy levels (measured using Likert scales) whether the narrative read was written in the first or third person. Kuzmičová et al. (2017: 143) examine a previously hypothesized link between literary fiction and increased trait empathy by analysing 'explicit markers of empathic response' in participants' post-reading justifications of passages they marked striking or evocative while reading either a short story high on foregrounding or a manipulated version of that story without limited foregrounding. They find that the non-literary version of the text with limited foregrounding elicited more empathetic elaborations than the literary version with foregrounding and thereby failed to confirm their hypothesis.

Fernandez-Quintanilla (2020) takes a more holistic approach to narrative features by combining analysis of point of view presentation, characters' discourse presentation, character's emotion presentation and characterization techniques into full stylistic analytical accounts of characters in two short stories. She then links these stylistic analyses to a thematic analysis of reader responses to these characters gathered in focus groups to understand the textual and readerly features that contribute to empathy. She concludes that in her case study, the cumulative effect of narrative devices related to the protagonist facilitates empathy for many readers unless their moral evaluation of the character prevents empathetic engagement. Fernandez-Quintanilla propels the stylistic exploration of narrative empathy forward by arguing that textual features contribute to narrative empathy cumulatively rather than in isolation. She also stresses that readerly factors such as moral evaluation and reader positioning should be considered key for shaping narrative empathy.

Capturing narrative empathy empirically using think-aloud protocols

Fernandez-Quintanilla (2020) is a strong example of the body of stylistic research that has, in recent years, started using qualitative reader response data to underpin their stylistic accounts of texts (Whiteley and Canning 2017). Her findings imply that anyone taking a stylistic approach to narrative empathy should consider methods and measures that allow for consideration of the cumulative effect of textual features as well as various readerly factors at the same time. Indeed, the demonstrated efficacy of using reader response data for studying reading experience from a stylistic perspective does not preclude careful consideration of the kinds of reader response data able to capture the complexity of the experience under investigation. Based on Whiteley and Canning's (2017) discussion of using reader response data for stylistic purposes, we highlight three concerns that inform the choice of method for studying the role of language in shaping narrative empathy.

When around the act of reading the data is captured

As for the stylistic studies that explore narrative empathy mentioned above, László and Smogyvári (2008), Van Lissa et al. (2016) and Fernandez-Quintanilla (2020) capture empathetic responses *after* reading the story. There are two problems with using post-reading responses to examine narrative empathy: such responses may include retrospective rationalizations based on reading the complete narrative or that participants may have forgotten their exact reasoning for marking a passage as striking or evocative (Short and Van Peer 1989: 25). By contrast, Kuzmičová et al. (2017) use a protocol developed by Sikora et al. (2011) in which participants highlight striking or evocative passages of the narrative *while* they read. Participants are then asked after reading the complete narrative to elaborate on three markings of their selection.[4] Though this protocol captures emotional enactment while reading, the after-reading elicitation of explanations for what participants marked as striking or evocative risks post-hoc rationalizations and forgetfulness too. Even if these weaknesses are mitigated by readers leaving physical reminders of their processing in the form of highlighter markings that may activate their memory, narrative empathy may only be captured in its pure form using alternative methodologies.[5]

The type of data required for analysis

Reader experiences may be accessed either through verbal or non-verbal data. Collecting verbal data requires participants to verbally communicate their experience, for example, in questionnaires, interviews or focus groups, using either open or closed questions. This type of data may be used to understand individual experiences from the participant perspective. Non-verbal data of experiences is captured using numerical measurements such as reading times or physiological instruments. Data of this kind is useful for studies that require consistent measurements of responses across all participants. When considering textual and readerly factors shaping narrative empathy, verbal data is likely to be more useful given its capacity to communicate details and nuance. Indeed, all studies in the section on analysing narrative empathy use verbal data exclusively.

The circumstances around data capture

Swann and Allington (2009) distinguish between two kinds of data collection circumstances: experimental and naturalistic approaches. Examples of experimental approaches include questionnaires or non-verbal reading time measurements, which are usually elicited in controlled environments and may result in artificialized reading behaviour due to the experimental protocols attached to participants' reading practice. By contrast, naturalistic approaches have readers engage in their usual reading behaviours in natural environments, without direct researcher engagement. Examples of such environments include reading groups and review platforms like *Goodreads*. Experimental collection practices tend to focus on pre-specified and controlled aspects of text processing, whereas naturalistic studies allow for nuanced understanding of 'the range of uses to which a particular text is put by particular readers in particular contexts' (Whiteley and Canning 2017: 77). Whether using experimental or naturalistic data, stylisticians relate reader interpretations and experiences to textual cues. Studies into how textual features shape narrative empathy may benefit from both experimental and naturalistic data, depending on whether the research question requires control over the contents of the data gathered.

We took these three considerations into account when we devised our methodology for Stradling and Pager-McClymont (2023). This study aimed to investigate the role of PF in shaping narrative empathy, and particularly how readers might draw on PF to empathize (see the section on analysing pathetic fallacy for how we defined and investigated PF). Answering these research questions required a detailed examination of readers' subjective reading experiences, which is why we deemed the elicitation of *verbal* reader response data most suitable. Furthermore, to counteract risks of post-reading rationalization or forgetfulness, we deemed it most appropriate to examine narrative empathy *while* reading. We thus opted for a think-aloud protocol, which presents narratives in successive passages to elicit reader responses to each paragraph in turn. Think-aloud protocols were previously used in spoken or written form by stylisticians to study, for instance, literary reception (Andringa 1990), meaning construction (Short and Van Peer 1989) or text-world processing (Norledge 2016). Such protocols are particularly powerful for linking specific experiences to specific paragraphs as well as for tracing reader experience across the whole reading process.

Think-aloud protocols may be considered both experimental and naturalistic in nature. Since formulating responses after reading each section is not typical of the natural reading process, reading while following a think-aloud protocol could lead to artificial experiences that do not resemble the natural reading experiences. However, as Norledge (2016: 66) argues, think-aloud protocols may be 'more or less naturalistic depending on the design of the study itself'. For example, the artificiality effect may be mitigated by presenting the complete narrative in its natural paragraphs, inclusive of paratextual features, and giving participants the freedom to read the narrative in their context of choice. In Stradling and Pager-McClymont (2023), we therefore presented participants with *The Flowers* in its natural paragraphs and including title, author and date of publication. We also refrained from limiting reader responses to just empathetic responses but asked participants to jot down per paragraph any impressions, experiences or memories salient to them.

The think-aloud protocol was followed by additional questions to prompt participant reflection on the story, the protagonist and the significance of the PF mappings (see Appendix 4.1 for the complete list of questions). Answers to these questions were meant to further nuance findings from participants' think-aloud responses. Participants were also asked about their empathetic disposition using questions from the Literary Response Questionnaire (Miall and Kuiken 1995) to check whether a lack of empathetic think-aloud responses may be due to participants' inability to empathize generally.

Analytical methodologies for PF perception and narrative empathy

Forty-two participants participated in our study. This section illustrates how we analysed their responses for perception of PF and empathetic engagement with the protagonist in the paragraphs of Walker's *The Flowers* (1973) that feature PF.

Analysing pathetic fallacy

We used Pager-McClymont's stylistically informed model (2021; 2022) to define PF and investigate it systematically in *The Flowers*. Pager-McClymont defines PF as 'a projection of emotions onto the surroundings by an animated entity. The emotions and animated entity in question can be featured implicitly or explicitly in the text' (2022: 435). This definition requires three criteria to be present in texts: presence of an animated entity, emotion and surroundings. Surroundings need to be described richly enough to allow readers to perceive the mirroring process of the emotions onto the surroundings. Pager-McClymont (2021: 176) argues that PF is a conceptual metaphor that is most often extended in texts ('megametaphor', Kövecses 2002: 51), and thus draws on Conceptual Metaphor Theory (Lakoff and Johnson 1980) as part of her model. As per Pager-McClymont's definition of PF, PF's conceptual master mapping can be identified as EMOTIONS ARE SURROUNDINGS. Because this mapping can encompass more specific mappings such as EMOTION IS COLOUR TONE such as GOOD IS LIGHT or BAD IS DARK (Forceville and Renckens 2013), it is thus a 'master metaphor' (Kövecses 2008: 382). Furthermore, texts containing PF are likely to feature the following linguistic indicators of PF: imagery (i.e. figures of speech), negation (syntactic, adverbial, morphological) and repetition (of lexis or syntax). Lastly, Pager-McClymont identifies at least five effects of PF in narratives:

1. communicating implicit emotions explicitly;
2. building of 'ambience' (Stockwell 2014), including suspense;
3. building characters (PF being an implicit cue of characterization in Culpeper's model (2001; see Pager-McClymont 2021: 297–8));
4. plot foreshadowing, which can generate suspense;
5. generating humour (Pager-McClymont 2023).

In this chapter we summarize how we analysed the presence of PF in *The Flowers* by drawing on Pager-McClymont's model. For more detailed stylistic analyses of *The Flowers*, see Stradling and Pager-McClymont (2023) and Pager-McClymont (2021: 281–3).

Walker's *The Flowers* (1973) is a short story about a Black girl named Myop, aged ten, picking flowers in the woods when she comes across a dead man. Although no year is specified, the story is set in the summer on a sharecropper's farm (post-American Civil War). PF is present in three distinct mappings:

1. GOOD IS LIGHT in paragraphs 1 and 2,
2. BAD IS DARK in paragraph 5,
3. EMOTIONAL CHANGE IS SEASONAL CHANGE in paragraph 8.

In paragraphs 1 and 2, Myop's joy and innocence can be inferred from the surroundings, as shown by the phrase: 'She felt light and good in the warm sun' (Walker 1973: 120). The light surroundings reflect Myop's positive emotional state, thus generating the PF mapping GOOD IS LIGHT. This builds Myop's character and creates a positive ambience to set the scene. In paragraph 5, Myop has reached a part of the forest she does not recognize and descriptions of the surroundings include 'the strangeness of the land', and 'the air was damp, the silence close and deep'. Myop's feelings of discomfort or fright can be inferred through her impression 'It seemed gloomy in the little cove', and 'Myop began to circle back to the house, back to the peacefulness of the morning' (Walker 1973: 120). This generates the PF mapping BAD IS DARK, stressing the foreboding aspect of the scene's ambience and foreshadowing a negative plot twist: Myop discovering a lynched man's corpse. In paragraph 8 (the story's final paragraph), the description (and arguably Myop) focuses on the frayed noose around the corpse's neck. The story ends with: 'Myop laid down her flowers. And the summer was over'. Here, PF's mapping is EMOTIONAL CHANGE IS SEASONAL CHANGE because Myop's childlike innocence is conveyed through the change of season, even though her action of laying down flowers (a symbolic act at a funeral) logically would not trigger a change from summer to autumn, which further characterizes Myop.

We also drew on Pager-McClymont's model to devise a method to observe readers' perception of PF, since no such method yet existed. To examine potential PF perception in online and offline processing separately, we differentiated between PF perception in think-aloud responses and prompt questions. Using a spreadsheet, we recorded mentions of any of PF's criteria, indicators or effects into dedicated columns and separated mentions before or after prompts by a dash. We also noted any pertinent specific mentions from the participants. If participants did not mention an element, we entered a slash. Table 4.1 below (adapted from Stradling and Pager-McClymont 2023) reflects the codes we used alongside points awarded for each PF element.

Table 4.1 Scoring system used to track participants' perception of PF

	Element of PF as per Pager-McClymont's model (2021, 2022)	Code used to log it in participants' reflections	Points awarded
Criteria of PF's definition	Animated entity (here human being)	H	1
	Surroundings	S	1
	Emotions	E	1
Linguistic indicators of PF	Imagery	I	1
	Negation	N	1
	Repetition	R	1
Effects of PF	Symbolism through Foreshadowing	S	1
	Communication of implicit emotion	I	1
	Characterization	C	1
	Building ambience	A	1
Scoring total maximum			**10**

Figures 4.1–4.4 provide screenshots that illustrate our recording process. Bolded text is relevant to our PF analysis, and text in italics refers to context or intertextuality.

The second column in all four figures shows the participants' total score of PF perception before and after prompt questions. These scores were calculated using the point system in Table 4.1. Table 4.2 below (adapted from Stradling and Pager-McClymont 2023) summarizes the scale of PF perception based on participants' scores.

Overall, participants who scored below 3 did not perceive PF, as the three criteria needed to fulfil PF's definition were not all accounted for. Participants who scored above 3 by discussing PF's effects or indicators but did not perceive the three criteria ultimately did not perceive PF due to its definition not being completed. This is the case for participant 1631471 in Figure 4.3: Myop's emotions are not discussed by the participant before or after prompt. As such, even though the participant scores 3 before prompt and 4 after, overall, they did not perceive PF. Participants who scored 3 by discussing all of PF's criteria had the potential to perceive PF. Moreover, participants who scored 4–5 by including PF's three criteria and one or two effects or indicators were noted to perceive PF. This is the case for participant 1841537 (Figure 4.1), who scored 4 before the

Random ID	PF SCORE /10	PF CRITERIA	PF INDICATORS	PF EFFECTS	SPECIFIC MENTION	"DEEPER SYMBOLIC MEANING"
1841537	4 BEFORE PROMPT 5 AFTER PROMPT	S H E	/	A – S		

Q8	Q9	Q10	Q11	Q12	Q13	Q14	Q15	Q25
Paragraph 1	Paragraph 2	Paragraph 3	Paragraph 4	Paragraph 5	Paragraph 6	Paragraph 7	Paragraph 8	**What do you think these sentences represent in the overall story?** *Myop laid down her flowers. And the summer was over.*
This made me think about spending early mornings in the countryside when I was little	**I could picture the setting in my mind.** This made me think about how life is easier when you are a child.	The image of the stream is the most vivid, it gives me an impression of summertime, of calm and silence	Curiosity, openness to nature	I'm starting to feel a little concerned and anxious for the setting in which she is.	I am feeling anxious and cannot understand what happened. I feel as if something horrible is about to be discovered	I am feeling horrified, as I understand she has found a dead body. Also, **she seems more curious rather than afraid, and this feels a little uncomfortable**	As before, I feel uncomfortable as she is interested in the rotten parts of the man's body and doesn't seem to be afraid or horrified.	These isolated sentences could represent her abandoning of childhood innocence, that she has somehow grown with that experience, but this thought has just come to me now, when I re-read these sentences.

Figure 4.1 Example 1 of participants' think-aloud responses.

Random ID	PF SCORE /10	PF CRITERIA	PF INDICATORS	PF EFFECTS	SPECIFIC MENTION
1757456	7 BEFORE PROMPT 8 AFTER PROMPT	E H S	I	A C S – I	FOCUSES ON IMAGERY + SYMBOL OF SEASON + "FORESHADOWING"

Q8	Q9	Q10	Q11	Q12	Q13	Q14	Q15	Q25
								What do you think these sentences represent in the overall story? *Myop laid down her flowers. And the summer was over.*
Paragraph 1	Paragraph 2	Paragraph 3	Paragraph 4	Paragraph 5	Paragraph 6	Paragraph 7	Paragraph 8	
Myop is introduced here and I am assuming she will be the main character. The verb skipped implies youth and happiness and coupled with the idea that every day is a "golden surprise" I get the impression that she is innocent and yet to be hardened by the real world, so I picture a child. The choice of food stuffs, "Corn, Cotton, Peanuts and squash" suggest a fertile home and	My mental images are resolving as we go through this second paragraph. We learn Myop is brown skinned, so I think Native American more likely than African American. We learn she "struck out" at chickens she liked, not that she didn't so she is associating with affection, a worrying combination, especially when in the previous paragraph we saw	Ok, so I am drawn to the word Sharecroppers. Are we in the South post-reconstruction? I am drawn to that setting because that is related to my area of study. It would also make sense with Alice Walker's position on civil rights. Also, sharecropping implies not a reservation so I am drawn away from Myop as Native American and now consider maybe she	Now we see threats introduced to the innocent world of the girl. Sure, we have "pretty flowers", "fragrant buds" and "sweet suds bush" all of which seem reassuring and her mother took her here so she has a positive connection to the place. But we have "snakes" now, so I am expecting a threat to rear up. Also the flowers are now "strange" so unfamiliar. I am pretty certain	Yep, I feel pretty happy here that my initial thoughts seem to be correct. We have growing threat ("Mile or more from home", "Strangeness of the land", "Gloomy", "Silence close and deep") so I am expecting a second character or possibly an animal to appear in the next paragraph, the antagonist of the tale. I'm remembering my first reading of Angela Carter's "The Bloody Chamber"	She found a corpse it seems. "He" is on the ground and she steps on his face and we are given no information for sound so I assume he is dead. She is unafraid of it, so with her earlier connotation of innocence she isn't really aware of death yet. "naked grin" is an odd phrase, threatening, sinister but also suggesting nothing hidden or held back. He is open to her interpretation.	Yep, a corpse. Happy I was right in that. There is something going on with scale here and that interests me. Walker describes him as "tall", "large", "big" and "long" and the girl seems small from her childlike traits. Not sure what, if anything that is about yet. I am also seeing the connection of late autumn from a paragraph or two back, harvest time from the beginning	Lynching. I am thinking of lynching here. We have the noose, the tree branch, the abandoned body. The single sentence paragraph is powerful as well "The summer was over" is suggesting her innocence has gone, the joy of her youth, the beauty of the long hot days fading to the upcoming darkness as the reality of the world washes over her in this moment. It was and rot,	That last sentence is the climax of the whole story. It's the end of her innocence, the death of her protected world view. Its deeply tragic. The laying down of her flowers is both a reflection of her shock as what she realizes she has stepped in and also the abandonment of her previous worldview.

(continued)

the word harvesting suggests Autumn so I see a reddish sky behind the girl, probably African American or Native American from the name, who I picture with plaited pigtails because my mind creates stereotypical images I guess. The two phrases that stand out for me in the passage as odd though are "jaws" and "nose twitch". There is something animalistic in those images to me (Nose twitch suggests rabbit, jaws a predator) so I am wondering if themes of predator and prey will matter here.	her connected to predator and prey. She could well be abuser because she was abused if this is foreshadowing. She is singing, reasserting the innocent and carefree joy she finds in the world. She is also lost in the moment, her world just where she is at that time. I am now fully expecting something awful to happen to her and while that will probably be sad, I am more curious of what Walker is doing here than worried or concerned.	is brown skinned because she is of mixed racial heritage? There is also an interesting image in "white bubbles disrupt the thin black scale" which I think is possibly metaphorical but I need to know more what is happening to put it in place correctly. I remain curious.	we have a tragedy coming up here. I am also now aware from the combination of a child "making her own path" and the fact she is only "vaguely" watching for threats that this may be a sort of reworking of Little Red Riding Hood and the connotations that carries. I remain curious to read on and see which of my suspicions are true.	for some reason at this point, probably from the Red Riding Hood vibe I am getting.
			suggesting the time of endings, of the passing of summer and the threat of oncoming winter. The progression from images of plenty and food at the beginning (Early autumn) to these connotations (Later autumn) is clever and I felt impressed at the skill employed here.	foreshadowed nicely, so I wasn't surprised by it, but I found it effective. I have quite a visceral reaction to the concept of lynching since it comes up fairly often in my research and I have read a lot of defenses of it from Ex-Confederates in the 1880's and 1890's so the anger I feel when faced with this historical practice flared when I saw the connection, but it didn't detract from the skill of the writer's build to the moment.

Figure 4.2 Example 2 of participants' think-aloud responses.

Random ID	PF SCORE /10	PF CRITERIA	PF INDICATORS	PF EFFECTS	SPECIFIC MENTION
1631471	3 BEFORE PROMPT 4 AFTER PROMPT	H S	/	C – S	"The surroundings in the story is allegorical."

	Q8	Q9	Q10	Q11	Q12	Q13	Q14	Q15	Q25
	Paragraph 1	Paragraph 2	Paragraph 3	Paragraph 4	Paragraph 5	Paragraph 6	Paragraph 7	Paragraph 8	What do you think these sentences represent in the overall story? *Myop laid down her flowers. And the summer was over.*
	I have never experienced such situation. so I cannot write much about this paragraph.	**Myop was a lovely girl and she deserve a good fortune.**	It seems that the story progress in the due course of time	She is growing day by day.	She has to realise the responsibility of being a girl.	**She started realising the responsibilities.**	I can imagine now the entry of a rough man.	Seasons passed in no hurry.	It is the period of transition.

Figure 4.3 Example 3 of participants' think-aloud responses.

Random ID	PF SCORE /10 9 BEFORE PROMPT	PF CRITERIA	PF INDICATORS	PF EFFECTS	SPECIFIC MENTION
1721594	9 BEFORE PROMPT	H S E	N I	A I C S	PERCEIVES SYMBOLISM OF PF AND USES METAPHOR TO DESCRIBE IT: "The realisation of how he died marks the end of her innocent enjoyment and, symbolically, her innocence more generally. Her lovely simple world is poisoned."

Q8	Q9	Q10	Q11	Q12	Q13	Q14	Q15	Q25
Paragraph 1	Paragraph 2	Paragraph 3	Paragraph 4	Paragraph 5	Paragraph 6	Paragraph 7	Paragraph 8	What do you think these sentences represent in the overall story? *Myop laid down her flowers. And the summer was over.*
This is an unfamiliar landscape to me. Sounds from a past period of history and in a place foreign to me. The person's name is also unfamiliar. Given what sounds quite a basic and relatively unappealing place (pigpen, smokehouse) the person seems particularly happy and content. Not sure why these things are a 'surprise' to her. Either she is very young or she had reason not to expect to share these experiences for some reason.	I grew up on a farm and the solitariness of children's play in general and that of a child not surrounded by other children is conjured. Odd to strike out at chickens she likes, maybe, but it doesn't sound malicious. She is self-absorbed and content, and living in a world of her own, as I did a lot as a child.	*Sounds like down South US*, as I had thought both from previous detail and the author. Sounds poor and scruffy, yet is described in a way that finds beauty in simple things. The child does not feel deprived or distressed and is captivated by small and simple things like the bubbles in the mud. I too enjoyed the natural world around my home as a child and was often alone in it	Again the simple countryside is made to sound fascinating and beautiful to the child and a source of nourishment, physical and psychological	This begins to sound a bit threatening, both the distance that she has travelled from home and the use of less attractive language such as 'damp', 'Silence close and deep'. It makes me think both of the distress of a child who realises they are lost and also of the possibility of some real danger to her	Confusing section. At first I thought her being 'lodged' was metaphorical but then it seemed she was really stepping on someone or something that looked like a person. A dead body? Disturbing.	Ok, a body. The description makes it clear he has been dead for a while, which somehow makes it less threatening, but it is an ugly image, especially the fact his head has been separated from his body. She seems more curious than afraid, as she pushes aside the leaves	*Ah, a lynching*. Odd that she seems calm enough to pick a rose and it is only when she sees the rope that 'the summer is over', as if his death alone wouldn't have been enough to distress her. Not clear if she knows at once what the significance of the rope is and that is enough to end her summer, or whether it will be when she tells her parents that the reality will become clear. A sense of her lost innocence. Nothing like this has happened to me, I'm glad to say, although we have all had events in our lives which cause us to lose our innocence, some welcome, some not	Her innocence is gone and all the joy she felt and the safety and familiarity of her surroundings and life were undermined by realising (or perhaps being reminded) that the world was not a safe place for people like her family. That the world was a cruel and violent place.

Figure 4.4 Example 4 of participants' think-aloud responses.

Table 4.2 Participant's scores of PF perception

PF scores (out of 10)	Composition of scores	Meaning of scores for PF	Participants' scores *before* prompts (/42)	Participants' scores *after* prompts (/42)
6+	three criteria and at least three effects and/or indicators	Interprets PF	18 = 43%	31 = 74%
4-5	three criteria and one or two effects and/or indicators	Perceives PF	14 = 33%	9 = 21%
3	three criteria but no effect or indicators	Potential to perceive PF	8 = 19%	1 = 0.2%
0-2	only two criteria or one criteria and one effect or indicator	PF is not perceived	2 = 5%	0 = 0%

prompt questions (PF's three criteria + the effect of building ambience), and 5 when they discussed the symbolic effect of the last paragraph after the prompts. Lastly, participants who scored above 6 both perceived and interpreted PF as they included all three PF criteria and at least three effects and/or indicators. This is the case for participants 1757456 (Figure 4.2) and 1721594 (Figure 4.4).

We also used this scoring system to analyse PF perception per paragraph. In that case, we were interested in PF perception per mapping across all participants to examine how often PF perception per mapping co-occurred with participant empathy.

Analysing narrative empathy

We used Fernandez-Quintanilla's (2018: 236) 'potential evidence-of-empathy codes' to examine think-aloud responses for evidence of empathy. Appendix 4.3 shows the codes, which are divided up into four categories with three sub-categories each. The first category of codes shows how participants adopt the perspective of a textual character for empathy. The second and third categories relate to how participants present their inferences of character's emotions (2) or other mental states (3). The fourth category's codes reflect participants' mental representations of character situation. While this table excludes explicit mentions of the word 'empathy', our analysis also includes these as potential evidence of empathetic engagement. We consider participants to have potentially empathized with Myop if their think-aloud response can be coded with one of the above categories. Responses coded with multiple codes indicate more evidence for empathetic engagement, and a potentially richer empathetic response.

To ensure that we coded the think-aloud data correctly according to Fernandez-Quintanilla's framework, we use the self-explanatory explanations she provides for each category (see Appendix 4.3). Nevertheless, the cursory nature of some think-aloud responses prevented straightforward application of the codes in some cases. For example, participant 1841537's response to Q11 'curiosity, openness in character' (Figure 4.1) is not elaborate enough to know whether these characteristics refer to Myop only or also were also shared by the participant. In such ambiguous cases, responses were never coded as evidence of empathy. Not all responses that mentioned participant feelings were marked as evidence of empathy either, usually due to a lack of clarity around these feelings being shared with Myop. For example, participant 1841537 (Figure 4.1) mentions 'starting to feel a little concerned and anxious' (Q12), 'feeling anxious' (Q13), 'feeling horrified' (Q14) and feeling 'uncomfortable' (Q15) but subsequently does not demonstrate having inferred Myop to have those feelings too. Furthermore, they describe their impressions of Myop's surroundings without any indication they vicariously share that experience of the setting (as in their responses to Q8 and Q9). Therefore, we did not mark any of participant 1841537's responses as containing evidence of empathy.

Some responses could not in its entirety be marked as displaying potential empathy either. Multiple of participant 1757456's responses exemplify such cases (Figure 4.2). Though their responses do not evidence they affectively share with Myop their impressions of Myop's settings, there are indications that they are drawing on intertextual references to empathetically attribute characteristics to Myop (per the attribution categories for emotions and mental states). For example, in response to Q11 they draw on the textual descriptions 'she made her own path' and 'vaguely keeping out for snakes' to integrate their knowledge of *Little Red Riding Hood* into their representation of Myop's circumstances. They thus show evidence of using their reader background to mentally attribute characteristics to Myop, which implies they understand her circumstances in affective detail. We, therefore, marked some of these responses as partly empathetic.

Responses were coded as wholly empathetic when most of the response could be coded as evidence for empathy or when responses could be coded with multiple categories. For example, participant 1721594's response to paragraph 2 (Q9, Figure 4.4) could be coded in its entirety with four categories. This participant attributes emotional experience ('she is content'), evaluates what experience is like (e.g. 'she is self-absorbed and content, living in a world of her own'), displays affective understanding (e.g. 'as I did a lot as a child') and attributes situational factors to Myop's experience (e.g. her living in a world of her own explains her contentment). The participant also seems to integrate Myop striking the chickens into this mental representation of a solitary child playing, showing that their past experience even allows them to share Myop's motivation for this action empathetically.

Once we identified which participant responses to which paragraphs exhibited evidence of empathy, we were able to consider whether participants empathized with Myop across the whole narrative. To do so, we looked at how many participants empathized with Myop per paragraph. Because we found that very few participants empathized with Myop in all eight paragraphs (see Stradling and Pager-McClymont 2023 for more detail), we subsequently assessed which paragraphs received the most empathetic responses. We found that fewer participants showed evidence of empathy towards the end of the narrative than at the beginning. As PF mappings are present throughout the narrative, we concluded that empathetic engagement fluctuated in response to the different paragraphs with different PF mappings.

The first factor we considered as a potential reason for this finding was perception of PF. We did so by relating the PF analysis described in the section on analysing pathetic fallacy to the three paragraphs with different PF mappings. We cross-tabulated these findings with our empathy analysis to examine whether participant empathy bore any relation to non-perception. Our findings, described in the section below, showed that this was not the case; overall participants show evidence of PF perception of each mapping, but perception does not automatically lead to empathy for Myop. These findings led us to return to the responses to paragraphs that explicitly featured a PF mapping to see whether these provided other reasons for either empathetic or non-empathetic responses. We looked at responses to paragraphs 1 and 2 for the GOOD IS LIGHT mapping, paragraph 5 for the BAD IS DARK mapping and paragraph 8 for the EMOTIONAL CHANGE IS SEASONAL CHANGE mapping. We were not just interested in whether participants empathized per mapping, but in the ways they verbally reported their empathy according to Fernandez-Quintanilla's categories. Therefore, we used the codes both to distinguish between empathetic and non-empathetic responses in relation to each paragraph, and to identify reasons for non-empathy as we considered why responses could not be coded as empathetic. Linking these analyses to the paragraphs that feature PF then helped us understand how the conceptual mappings inherent in this narrative's instantiations of PF might shape empathy.

Summary of data

Stradling and Pager-McClymont's (2023) main findings emanate from the integration of the PF recognition analysis described in section on analysing pathetic fallacy and the evidence for empathy analysis described in section on analysing narrative empathy. By showing how many participants that displayed evidence of empathy or lack thereof with any of the three PF mappings also perceived PF, Table 4.3 helps us to consider whether perception of PF can explain the discrepancy in empathetic engagement per PF mapping observed in our empathy analysis. For the GOOD IS LIGHT mapping we separated the number of participants per perception category for paragraphs 1 and 2, while we also include the number of *unique* participants per category to indicate how many participants perceived the PF mapping either in paragraph 1 or 2. In cases of non-perception or no mention, we only counted those participants who did not perceive or did not mention the mapping across both paragraphs. We included two columns for the EMOTIONAL CHANGE IS SEASONAL CHANGE PF mapping: one reflecting PF perception in our think-aloud data, and one reflecting PF perception in prompt question Q25. We observed in our PF analysis that seventeen participants who did not mention PF in their think-aloud responses did reflect on the effect of PF in their response to Q25, implying they perceived PF while reading despite their think-aloud responses not displaying it. If we take responses to Q25 into account, Table 4.3 shows that the lack of perception or mention of PF across the mappings remains relatively stable. Although the number of participants who do not mention PF is higher for the EMOTIONAL CHANGE IS SEASONAL CHANGE mapping, this can be explained by participant focus on narrative interpretation resulting from this mapping being featured in the final paragraph of *The Flowers*.

Based on Table 4.3, we argue that our study does not provide conclusive evidence that PF perception leads to empathetic engagement. If that were the case, fewer participants who did not empathize would also not have perceived PF. Instead, we can observe that most participants perceived PF but did not always empathize. Specifically, we see the number of participants who empathize decreases per PF mapping, while the number of participants who do not perceive PF remains relatively low across the mappings. That leads us to conclude that the role of PF on empathy is mediated by the kind of PF mapping used; the specifics of each mapping seem to impact how many participants empathize.

We subsequently investigated how the different PF mappings might lead participants to empathize by considering how their responses showed that participants drew on each mapping to process Myop's experience. We find that participants who empathize with Myop in paragraphs 1 and/or 2, which include the GOOD IS LIGHT mapping, draw on personal past experiences of similar situations; their responses were coded principally with situation codes alongside emotion codes. For instance, participant 1721594 (Q9, Figure 4.4) draws on their experience of growing up on a farm to characterize Myop's mental states

Table 4.3 Perception of PF and evidence of empathy in think-aloud responses cross-tabulated

PF Mapping		Empathy			No empathy		
		Perceived	Not perceived	No mention	Perceived	Not perceived	No mention
GOOD IS LIGHT ($1/$2)	$1	21/42 = 50% | 28/42 = 67%	1/42 = 2% | 2/42 = 5%	5/42 = 12% | 2/42 = 5%	9/42 = 21% | 7/42 = 16%	1/42 = 2% | 0/42 = 0%	5/42 = 12% | 2/42 = 5%
	$2	22/42 = 52%	3/42 = 7% | 2/42 = 5%	2/42 = 7%	7/42 = 17%	3/42 = 7%	5/42 = 12%
BAD IS DARK ($5)		15/42 = 36%	0/42 = 0%	3/42 = 7%	20/42 = 48%	0/42 = 0%	4/42 = 10%
EMOTIONAL CHANGE IS SEASONAL CHANGE ($8)		5/42 = 12%	0/42 = 0%	3/42 = 7%	10/42 = 24%	0/42 = 0%	24/42 = 57%
EMOTIONAL CHANGE IS SEASONAL CHANGE ($8) based on Q25*		7/42 = 16%	1/42 = 2%		26/42 = 62%	8/42 = 19%	

*Because Q18 indirectly required participants to reflect on the effects of this PF mapping, not mentioning PF's effects or the correlation between the PF's domains means participants did not perceive this PF mapping.

as 'self-absorbed and content ... as I did a lot as a child'. Where think-aloud responses cannot be coded using Fernandez-Quintanilla's codes, participants instead just describe Myop's situation or mention personal emotions not shared with Myop. Participant 1757456's (Q9, Figure 4.2) response exemplifies responses that express emotions not shared with Myop; they verbalize processing Myop as 'brown skinned', 'abuser' and 'singing' while considering how her striking chickens worries them. Our analysis concludes that the GOOD IS LIGHT mapping affords empathy by inviting readers to use positive childhood experiences to create a mental representation of Myop. Reasons for a lack of empathy seem to lie in participants' failure to relate their understanding of Myop to their self.

We find a slightly different pattern for the BAD IS DARK mapping. Participants who show evidence of empathy do so by verbalizing it using imagined scenarios associated with danger, often alongside verbalizations that may be coded with emotion codes. For example, participant 1363237 (Figure 4.4) attributes situational factors to Myop's mental states (e.g. 'the distance from home') by drawing on what a child who is lost or facing danger (which is an imagined scenario) may feel. Participants who do not show evidence of empathy instead show evidence of fear not shared with Myop, or consider the PF mapping to contribute to plot progression (particularly foreboding) rather than Myop's experiences here. Participant 1841537's response to paragraph 5 (Q11, Figure 4.1) exemplifies this: they feel 'anxious for the setting in which she [Myop] is' (sympathetic response) rather than anxious with her (empathetic response). Therefore, this PF mapping seems to invite mental representations of Myop built on past experiences of danger, though empathetic responses may only be reported if participants do not (also) have related experiences. Our analysis of non-empathetic responses indicates that for some participants fear for Myop (sympathy) rather than with her (empathy) may be a more salient experience, while for others a focus on plot progression may mean they do not report empathy.

Our analysis of responses to paragraph 8's description of Myop using the EMOTIONAL CHANGE IS SEASONAL CHANGE mapping demonstrates that this PF mapping invites interpretations of Myop's circumstances, which in turn impact empathy or lack thereof. For instance, as participant 1631471 (Q15, Figure 4.3) only records their interpretation ('seasons passed in no hurry') without referring to their or Myop's mind, their response cannot be coded as empathetic. Other participants interpret Myop to have lost her innocence, and either mention that interpretation without accompanying affective experiences or empathize with Myop's inferred loss of innocence. Another way participants draw on this paragraph's PF mapping is to infer Myop's mental states at the sight of the lynched man, which very few then empathize with. Participant 1841537's response (Q15,

Figure 4.1) is representative of this: their feeling 'uncomfortable as she [Myop] is interested in the rotten parts and doesn't seem to be afraid or horrified' lacks empathy as their negative affective state of discomfort is not shared with Myop.

Affordances of think-aloud data

We demonstrate that think-aloud protocols allow for the analysis of PF perception in readers, particularly to observe per paragraph whether participants mentioned any of PF's elements (criteria, effects, indicators) in Pager-McClymont's model. In combination with our newly created scoring system (Table 4.1), we could determine which elements of PF participants most commented on: here, criteria of surroundings and the foreshadowing effect. This unique combination of methods thereby provided a concrete illustration of Pager-McClymont's theoretical model of PF, as it allowed us to confirm her model's hypotheses that foreshadowing, characterization and building ambience are effects of PF readers recognize while reading. Nevertheless, we also found that the open-ended nature of a think-aloud protocol limited PF perception analysis; in the case of the novel mapping EMOTIONAL CHANGE IS SEASONAL CHANGE responses to prompt questions were required to observe participant PF perception, since participants chose to focus on other aspects of their experience in their think-aloud responses.

We can also establish methodological implications from our study for using Fernandez-Quintanilla's codes to analyse think-aloud data. Though the codes were originally developed to analyse rich reader data from focus groups, we find that they are well-suited to think-aloud data analysis. Think-aloud data combined with Fernandez-Quintanilla's codes helped us identify whether and where in a narrative participants might be likely to empathize, particularly because the data is linked to specific paragraphs. It also allowed us to identify both textual reasons (in this case, how participants drew on PF mappings) and readerly reasons (such as (lack of) shared experiences or focus on textual processing over felt experience) for why participants might (not) empathize.

However, we also found that think-aloud responses may not capture all experiences pertinent to the study of empathy. Just as participants did not always mention PF elements in their think-aloud responses (while they did in response to prompt questions), so empathetic experiences may not always be reported. Indeed, think-aloud responses only capture those aspects of the reading experience that are salient to participants. Relatedly, unlike focus group data, the think-aloud protocol does not allow for further explication of responses when required for analysis, meaning that inherent in the analysis is a level of speculation about what participants might have meant. Nevertheless, we have shown in sections above that analysis of evidence for empathy and reasons for empathetic engagement is still possible, and that we could still observe empathy frequently. With the possibility to link this to specific places in the narrative, our think-aloud data enabled us to examine where in *The Flowers* empathetic engagement may be effectively elicited across many participants.

Lastly, our study shows that think-aloud data is suited to the analysis of multiple aspects of the reader experience at the same time. In this case, the data could capture both PF perception and evidence of empathetic engagement and allowed for analysis of these two aspects of reading experience. We could also track the progress of these experiences across the whole narrative, as well as link specific PF mappings to the experiences evoked by them. Being able to accomplish all this with one protocol may decrease the length of data collection processes, which benefits controlling respondent fatigue. It also means future studies do not require re-reading of a narrative to capture various textual and readerly aspects of the reading experience via different measurements, which decreases the risk of gathering online responses impacted by substantial reflection. Think-aloud protocols thus afford the more straightforward study of textual contributions to complex reader experiences like empathy.

Conclusion

To conclude, this chapter demonstrates the value of a single think-aloud protocol for the study of a specific conceptual metaphor (PF) and narrative empathy. Elicitation of this online reading data enabled examination of the experiential impact of this conceptual metaphor mapping by allowing us to 1) identify how readers use the textual trigger to form mental representations and 2) identify their experiential impact. Furthermore, this dataset afforded the pioneering analysis of readers' perception of PF and the first application of Fernandez-Quintanilla's framework to online processing data. Crucially, being able to link our analyses of participant responses to individual paragraphs allowed for observing divergences in empathetic engagement shaped by *The Flowers*'s three different PF mappings.

This chapter underscores the increasing use of reader response data in stylistic research as a positive development. Increased use of reader response data impacts the evolution of improved data collection protocols and analytical frameworks. As we have shown, opportune reader response data has the potential to support existing and develop new research claims if combined with fitting analytical tools, including making theoretical models more concrete.

Notes

1 Following Castaglione (2017), 'online' processing refers to readerly processing while reading. This may be compared to offline processing, which refers to processing after reading.
2 This contrasts with research examining the long-term effect of reading texts on empathy as a character trait, which considers textual features as contributing to readers' social cognitive abilities beyond immediate empathetic engagement with characters.
3 Since the time of writing, a special issue on the stylistic study of narrative empathy in the *Journal of Literary Semantics* (Fernandez-Quintanilla and Stradling 2023) was published that includes some empirical investigations of empathy from a stylistic perpsective.
4 Kuzmičová et al. additionally used post-reading Likert scales about empathy-related phenomena like transportation and narrative engagement, which they linked to participants' elaborations.
5 Nevertheless, capturing narrative empathy in its pure form may not currently be empirically possible, since narrative empathy often co-occurs with other emotional and affective responses.

References

Andringa, E. (1990), 'Verbal data on literary understanding: A proposal for protocol analysis on two levels', *Poetics*, 19: 231–57.
Castiglione, D. (2017), 'Difficult poetry processing: Reading times and the narrativity hypothesis', *Language and Literature*, 26 (2): 99–121.
Culpeper, J. (2001), *Language and Characterisation: People in Plays and Other Texts*, London: Longman.
Fernandez-Quintanilla, C. (2018), 'Language and narrative empathy: An empirical stylistic approach to readers' engagement with characters', Doctoral Dissertation, Lancaster University, UK.
Fernandez-Quintanilla, C. (2020), 'Textual and reader factors in narrative empathy: An empirical reader response study using focus groups', *Language and Literature*, 29 (2): 124–46. https://doi.org/10.1177/0963947020927134 (accessed 12 January 2023).
Fernandez-Quintanilla, C. and F. Stradling (eds), (2023), 'Stylistic approaches to narrative empathy [Special Issue]', *Journal of Literary Semantics*, 52 (2).
Forceville, C. J. and T. Renckens (2013), 'The good is light and bad is dark metaphor in feature films', *Metaphor and the Social World*, 3 (2): 160–79. https://doi.org/10.1075/msw.3.2.03for (accessed 19 December 2022).
Keen, S. (2006), 'A theory of narrative empathy', *Narrative*, 14 (3): 207–36.
Kövecses, Z. (2002), *Metaphor: A Practical Introduction*, Oxford: Oxford University Press.
Kövecses, Z. (2008), 'Metaphor and emotion', in R. W. Gibbs (ed.), *The Cambridge Handbook of Metaphor and Thought*, 380–96. Cambridge: Cambridge University Press.

Kuzmičová, A., A. Mangen, H. Støle and A. C. Begnum (2017), 'Literature and readers' empathy: A qualitative text manipulation study', *Language and Literature*, 26 (2): 137-52.

Lakoff, G. and M. Johnson (1980), *Metaphors We Live By*, Chicago, IL: University of Chicago Press.

László, J. and I. Smogyvári (2008), 'Narrative empathy and inter-group relations', in S. Zyngier, M. Bortolussi, A. Chesnokova and J. Auracher (eds), *Directions in Empirical Literary Studies: In Honour of Willie van Peer*, 113-25. Amsterdam, Philadelphia: John Benjamins Publishing Company.

Miall, D. and D. Kuiken (1995), 'Aspects of literary response: A new questionnaire', *Research in the Teaching of English*, 29 (1): 37-58.

Norledge, J. (2016), 'Reading the dystopian short story', Doctoral Dissertation, University of Sheffield, UK.

Pager-McClymont, K. (2021), 'Communicating emotions through surroundings: A stylistic model of pathetic fallacy', Doctoral Dissertation, University of Huddersfield, UK.

Pager-McClymont, K. (2022), 'Linking emotions to surroundings: A stylistic model of pathetic fallacy', *Language and Literature*, 31 (3): 428-54. https://doi.org/10.1177/09639470221106021 (accessed 12 January 2023).

Pager-McClymont, K. (2023), '"The thunder rolls and the lightning strikes": Pathetic fallacy as a multimodal metaphor', *Anglica Wratislaviensia*, 61 (1): 53-75. https://doi.org/10.19195/0301-7966.61.1.4

Short, M. and W. Van Peer (1989), 'Accident! stylisticians evaluate: Aims and methods of stylistic analysis', in M. Short (ed.), *Reading, Analysing and Teaching Literature*, 22-71. London: Longman.

Sikora, S., D. Kuiken and D.S. Miall (2011), 'Expressive reading: A phenomenological study of readers' experience of Coleridge's The Rime of the Ancient Mariner', *Psychology of Aesthetics, Creativity, and the Arts*, 5 (3): 258-68.

Stockwell, P. (2014), 'Atmosphere and tone', in P. Stockwell and S. Whiteley (eds), *The Cambridge Handbook of Stylistics*, 360-74. Cambridge: Cambridge University Press.

Stradling, F. and K. Pager-McClymont (2023), 'The role of pathetic fallacy in shaping narrative empathy', *Journal of Literary Semantics*, 52 (2): 123-43.

Swann, J. and D. Allington (2009), 'Reading groups and the language of literary texts: A case study in social reading', *Language and Literature*, 18 (3): 247-64.

van Lissa, C. J., M. Caracciolo, T. van Duuren and B. van Leuveren (2016), 'Difficult empathy: The effect of narrative perspective on readers' engagement with a first-person narrator', *DIEGESIS. Interdisciplinary E-Journal for Narrative Research/ Interdisziplinäres E-Journal Für Erzählforschung*, 5 (1): 43-63.

Walker, A. (1973), *'The Flowers' in Love and Trouble: Stories of Black Women*, San Diego, CA: Harcourt Brace Jovanovich.

Whiteley, S. and P. Canning (2017), 'Reader response research in stylistics', *Language and Literature*, 26 (2): 71-87.

Appendices

Appendix 4.1 List of survey questions

Questions number	Type of question	Survey questions
1. – 6.	Multiple choice	Information Sheet and Consent Form Questions (Yes/No)
7.	Open Textbox	Please read the paragraph below. Please jot down in full sentences in the box below each paragraph any feelings, thoughts, images, impressions, or memories that come into your mind as you read each paragraph. You can write as much as you would like to and in whatever form you like. Short notes and bullet points are just as fine as full sentences. [**Title, Author, Date of Publication inserted**].
8.	Open textbox	Same as question 7. [**paragraph 1 of story inserted**].
9.	Open textbox	Same as question 7. [**paragraph 2 of story inserted**].
10.	Open textbox	Same as question 7. [**paragraph 3 of story inserted**].
11.	Open textbox	Same as question 7. [**paragraph 4 of story inserted**].
12.	Open textbox	Same as question 7. [**paragraph 5 of story inserted**].
13.	Open textbox	Same as question 7. [**paragraph 6 of story inserted**].
14.	Open textbox	Same as question 7. [**paragraph 7 of story inserted**].
15.	Open textbox	Same as question 7. [**paragraph 8 of story inserted**].
16.	Open textbox	Summarise the story in your own words.
17.	Open textbox	How do you feel after reading the story?
18.	Open textbox	How do you think Myop feels at the end of the story?
19.	Open textbox	Describe Myop's character in a few words.
20. A	Multiple choice	Did Myop's character evolve for you throughout the story? Yes/No
20. B	Open textbox	If YES, how did Myop's character evolve for you throughout the story? /If NO, why do you think Myop did not change throughout the story?
21.	Open textbox	If you ever were in Myop's situation, how do you think you would react?

22.	Open textbox	Describe the passage of the story you find the most striking and why.
23.	Open textbox	What do you think of the surroundings in the overall story?
24.	Open textbox	What do you think these sentences represent in the overall story? "It seemed gloomy in the little cove in which she found herself. The air was damp, the silence close and deep."
25.	Open textbox	What do you think these sentences represent in the overall story? "Myop laid down her flowers. And the summer was over."
26.	Open textbox	Why do you think the story is entitled "The Flowers"?
27.	Matrix table	Read each statement carefully. Then rate the extent to which the statement is true of you by clicking one circle per question. *See appendix 2.*
28.	Open textbox	**[Full short story included]** Are there any elements from the text that helped you visualise how Myop might be feeling? Please list them.
29.	Open textbox	Do you have any further comments? (about the text, your reading experience, or this study)
30.	Multiple choice	Have you read this story before? Yes/No
31.	Multiple choice	What is your gender? • Male (including transgender men) • Female (including transgender women) • Non-binary • Gender-fluid • Agender • Prefer not to say
32.	Multiple choice	What is your ethnicity? • White • Asian • Black (African or Caribbean) • Arab • Latino • Other • Prefer not to say
33.	Multiple choice	What is your educational background? • GCSE or equivalent • A Levels or equivalent • Undergraduate degree • Postgraduate degree • Other • Prefer not to say
34. A	Multiple choice	Did you encounter any difficulties in answering this questionnaire? Yes/No
34. B	Open textbox	If YES, could you explain what difficulties you encountered?
35.	Open textbox	What did you think of this questionnaire? Write as much as you like about what you liked or disliked, and feel free to provide ideas for improvement.

Appendix 4.2 Matrix table for survey question 20

Stimuli	Not at all true - false	Slightly true	Moderately true	Quite true	Extremely true
Sometimes I feel like I've almost "become" a character I've read about in fiction.					
I sometimes have imaginary dialogues with people in fiction.					
When I read fiction, I often think about myself as one of the people in the story.					
I sometimes wonder whether I have really experienced something or whether I have read about it in a book.					
I actively try to project myself into the role of fictional characters, almost as if I were preparing to act in a play.					
Sometimes characters in stories almost become like real people in my life.					
After reading a novel or story that I enjoyed, I continue to wonder about the characters almost as though they were real people.					

Appendix 4.3 Fernandez-Quintanilla's 'implicit-evidence-for-empathy codes' (2018: 236–7)

	CODE	DESCRIPTION
Perspective	*Character-oriented perspective taking*	Participants imaginatively adopt characters' viewpoint and focus on characters' inner states and circumstances (rather than self-orientedly imagining themselves in their situation)
	Attribution of speech/thought_ affiliation	Participants verbally articulate characters' speech/thought in direct form. This may suggest that participants simulate characters' mental activity. This is seen in sudden shifts to the 1st person pronoun.
	Pronoun use/shift	Readers' pronoun use shows differences between (i) talking about characters in the 2nd or 3rd person from an observer position and (ii) the verbal simulation or enactment of characters' experience in the 1st or 2nd person, where readers suddenly impersonate characters. Pronoun shifts can be accounted for in terms of a tension: when readers enact a character's consciousness a tension is created between the reader's simulation of the experience in the 1st person and the reader's attribution of the experience to the character in the 2nd or 3rd person (Caracciolo 2014: 110)
Emotions	*Attribution of emotional experience*	Participants attribute specific emotional states to characters. They spell out the emotional implications of story-world events; that is, what characters are likely to feel as a result of the story-world events (i.e. *pain, anguish*)
	Evaluation of what experience is like	Sometimes the attribution of emotional experiences (above) is coupled with evaluative expressions. This explicit element of evaluation indicates degrees of how distressing and undesirable the characters' emotional experience is (i.e. *the worst*)
	Affective understanding	Display of understanding of the character's emotional states based on first-hand experience: readers claim to have first-hand knowledge or experience of a similar situation and, as a result, they verbalize what the experience must be like for characters. These displays of understanding based on similarity of experience can suggest, as noted by Kuroshima and Iwata (2016), (i) affiliation with the target's stance towards the experience, (ii) understanding of the nature of the experience and its meaning (i.e. what the experience is like), and (iii) a congruent affective stance (i.e. potentially shared feelings)

	CODE	DESCRIPTION
Other mental states	*Attribution of thought processes*	Thought processes are attributed to characters
	Attribution of values and beliefs	Values and beliefs are attributed to characters
	Attribution of goals and needs	Goals and needs are attributed to characters' situation
Situation	*Attribution of situation*	Participants spell out characteristics of the situation characters are going through
	Imagined scenario	Participants describe a scenario parallel to the events undergone by characters, and they vividly depict the details of the situation, thus suggesting understanding and a potential projection into characters' situation
	Attribution of situational factors	Participants attribute situational forces to characters' actions and circumstances; that is, they provide contextual explanations for characters' behaviour.

5

Reader's Reactions to Descriptions of Landscape in Polish Translations of *Anne of Green Gables*

Beata Piecychna

Introduction

While in the last decades, general translation studies, perceived as an interdiscipline, has ventured beyond the linguistic and cultural perspectives into domains more connected with cognitive science, literary translation scholars, as opposed to literary scholars (see e.g. Semino and Culpeper 2002; Richardson 2010; Vermeule 2010; Zunshine 2010; Easterlin 2012; Bruhn and Wehrs 2013; Jaén and Simon 2013; Nikolajeva 2014; Zunshine 2015; Burke and Troscianko 2016), have not manifested their interest in developing literary translation studies as aligned more with cognitive approaches, for example, with the theory of mind, cognitive historicism or embodiment, to name just a few domains within the cognitive branch of contemporary literary studies.

Likewise, landscape studies, or, to put it more generally, the spatial turn, has not been explored by translation researchers in detail. Although Kershaw and Saldanha (2013: 136) are generally right when they claim that 'the notion of space has acquired a new relevance in translation studies', research on the subject has been rather restricted to limited and dispersed aspects of the category under focus (see e.g. Schulte 2001; Cronin 2003; Simon 2012; Italiano 2016; Farahzad and Ehteshami 2019).

Themes which are oftentimes discussed by translation scholars as 'spatial' trajectories of the translation process and product relate to, *inter alia*, the geopolitical and geographical transformation of cities, or other geopolitical, geographical and cultural zones. This process is frequently seen as a type of translation (2008; 2015a; 2015b; 2018; von Flotow and Nischik 2007; Cronin and Simon 2014), as translation treated as a specific space of cultural influences within a society (Baldo 2013; Buffery 2013; Hephzibah 2013; Marinetti and Rose 2013; Skomorokhova 2013; Torresi 2013), as translation seen from the perspective of geographical metaphors (Stecconi 2010) and as a translational indicator of globalization processes, also in the context of the book market (see e.g. Cronin 1996; 2000; 2003). Landscape in the studies mentioned above is treated specifically as a metaphorical and discursive domain referring to the notion of cultural and social translation, and not as a legitimate part of the plot of the literary text. Papers tackling the problem of translating descriptions of landscape *per se* are rarely published (see e.g. Williams and Marinkova 2015).

To the best of my knowledge, no studies have been published so far which would be conducted with the aim of analysing fragments of rendered descriptions of the landscape from the perspective of embodied aesthetics. It might be tentatively concluded that the cognitive component, despite its being an increasingly important element within a more empirical branch of translation studies, has not received critical attention among literary translation scholars yet. This chapter aims to fill this research gap. To that end, the main focus will be placed specifically on the ways *Anne of Green Gables* evokes the sense of landscape in the projected reader, or, in other words, on the way the notion of a given place is *simulated and aesthetically experienced* in the mind of the implied reader. Taking as a point of departure the claim that landscapes and their perception are developed in the social context, and thus they are modified together with cultural and historical changes, I intend to demonstrate not only how a selected description of landscape presented in the source text changes culturally and historically in their respective retranslations, but also how such changes might influence the reader's cognitive experience with the narrative. The claim is strengthened by the idea that cognition and culture are inextricably intertwined, with the agency of the author and of the translator playing here a fundamental role. In this way, I follow Mary Thomas Crane's (2001) postulates, and after her, I posit that 'cognitive theory offers new and more sophisticated ways to conceive of authorship and therefore offers new ways to read texts as products of a thinking author engaged with a physical environment and a culture' (2001: 4). By treating the author's and the translator's 'material existence' (2001), I intend to dig deeper into their ways of depicting landscape, juxtaposing it against the backdrop of the socio-cultural conditions of their times.

Assuming that landscapes are culturally, socially and historically determined, we should expect them to evoke different cognitive experiences and manifest themselves differently throughout the narrative depending on the author of the text and the times during which s/he lived. Thus in the case of retranslations, or translation series, aesthetic experience as connected with the perception of the landscape created by the successive translators might be evoked differently. Accordingly, this chapter might fill at least two prominent research gaps: 1) within the area of translation criticism and space and 2) within the area of cognitive translation criticism.

Although this has not been explicitly stated in the title of the article, this chapter employs the methodological paradigm of translational stylistics in that it attempts to analyse the impact of the translator's style, including particular words and phrases used in the target text, on the projected reader's reaction to the text they are reading (for more see Boase-Beier 2006). In this chapter then, my main interest will be placed on *how it is possible* for the text to mean, rather than what exactly the text means (see Boase-Beier 2014). And in harmony with Boase-Beier's (2006) views on studying style in translation, the focus will be placed on the area of the translator's linguistic choices and, consequently, on the relationship of these choices with a cultural, social and historical context in which the discussed translations were produced.

Embodied aesthetics: A brief overview

The embodiment paradigm is an interdisciplinary and multiperspective endeavour, and this interdisciplinarity and multiperspectivity should be understood as a twofold extension, that is, within the purview of making use of the latest findings in a wide variety of scientific disciplines, including, but not limited, to, psychology, philosophy, neuroscience, information technology, biology, linguistics, cognitive science and artificial intelligence, and within the process of analysing a plethora of aspects comprising human cognition, for instance language comprehension, making judgments and decisions, solving problems, understanding and reasoning, memory, learning, etc. The most fundamental assumption of the embodiment paradigm is that mental actions and processes of understanding, reasoning about and experiencing the world rest on the characteristics of the body, in particular on its sensorimotor and perceptual systems. A fundamental part of the embodiment paradigm is the embodied simulation hypothesis which states that people understand language by simulating what it would be like to experience the same event/action/state which they only observe or about which they read/hear (for more, see, e.g. Bergen and Wheeler 2010; Bergen 2012; Gallese 2012; Zwaan and Pecher 2012; Kok and Cienki 2017).

The mechanism of embodied simulation is powerfully influenced by the context (be it political, historical, cultural or social) in which the individual is embedded. Thus simulations are by necessity individualized and differentiated: 'Mirror neurons and embodied simulation do not consist of stereotyped and undifferentiated responses. They are both context-dependent and idiosyncratically linked to individuals' personal historical, social and biological identity' (Gallese 2019: 124). In the context of descriptions of landscape (itself being a specific form of a cultural artefact), one could undermine that bodily responses experienced by both writers (including translators) and readers are influenced by more abstract context-specific entities (politics, culture, society) within which writers, translators and readers (as individuals having their own history) function.

Interestingly enough, embodied simulation also takes place in the case of the reading experience of literary fiction. As Gallese rightly points out, 'Our relationship with fictional worlds is double-edged: on the one hand, we pretend them to be true, while, on the other, we are fully aware they are not' (2017: 47). The author explains this mechanism by referring to the fact that the reader distances himself/herself from 'the external world' (2017). Put simply, the rationale behind those studies is that when individual reads fiction, they mentally simulate the protagonist's actions, sensations, feelings and perceptions as if the individual experienced the same phenomena themselves. The author (2019: 116) also underlines that simulationism occurs in the case of visual imageries. As Gallese says, '[v]ision is a complex experience, intrinsically synaesthetic, that is made of attributes that largely exceed the mere transposition in visual coordinates of what we experience any time we lay our eyes on something' (2019: 118).

This relational quality makes it legitimate to claim that when experiencing a landscape as depicted in fiction, we also simulate what it would be like to sense such spatial entity in all its facets and dimensions through sensory modalities at hand. The perception of the image of landscape results from our constructing an experience by means of our sensory and motor systems (Gallese 2019: 117). All the simulations triggered by viewing and experiencing a landscape (including emotions, feelings, perceptions, etc.) can be classified without too gross a generalization as *embodied aesthetics*. For as Gallese (2017: 44) puts it, 'embodied simulation is also triggered during the experience of spatiality around our body and during the contemplation of objects'. As Gallese (2017) further claims, the latest findings in neuroscience show that what we see does not only encompass the visual but is composed of our imagination, memories, sensorimotor capacities of our body, emotions, feelings, senses, etc.

Finally, a few remarks should be made in regard to the rationale behind deploying the theoretical and methodological dimension of embodied simulation within the field of aesthetics and aesthetic experience. As Gallese (2017: 45) asserts, a vocal proponent of using embodied simulation as an approach to understanding the specificity of humans' encounters with images, claims, 'the heuristic value' of embodied simulation can be discerned within the area of aesthetics, in particular experimental aesthetics, in two ways.

First, as Gallese (2017) underlines, the relevance of embodied simulation to aesthetics can be seen in the fact that due to the Mirror Mechanism specific to the human brain, the individual experiences pieces of art, in particular images, in the bodily format, that is because of bodily feelings activated in response to looking at images. Second, the relevance could also be noted with regard to the way of creating the image and the recipient's bodily reaction to it in the form of simulating the author's creative process (2017).

And, as is commonly known, Montgomery made landscape one of the most important elements of her storyworlds. It does not only constitute a background for protagonists' actions, feelings and reflections but, first and foremost, provides an *interpretative force* and source of implied reader's aesthetic experience. And I posit that this experience is happening in and through the bodily format.

Methodology

For the purposes of this chapter, certain foregrounding features will be employed. Due to space constraints, only those foregrounding items will be referred to, which occur at a lexical level. This decision could be motivated by the fact that descriptions of landscape can be best addressed from the perspective of concrete lexical items, themselves being embedded in certain types of (ideologized) frames, which help the reader to visualize and simulate scenes about which they are reading as based on the embodied reality in which they function. Therefore, in this analysis, one of the most fundamental assumptions of Marvin Minsky's frame system theory (1988) will be used. According to Minsky (1988: 244), 'each perceptual experience activates some structure that we'll call *frames* – structures we've acquired in the course of previous experience ... each representing some stereotyped situation like … being in a certain kind of room'. In other words, in this chapter the claim is that the reader understands, interprets and *senses* a given spatial description based on their previous (embodied) encounters with the spatial and (geo)cultural in which they function. Besides, the assumption, in harmony with the main tenets of the embodiment paradigm, is that the implied reader simulates what it would be like to aesthetically experience that which Anne Shirley, the protagonist, sensed, saw and felt while seeing her new home for the first time.

In order to analyse chosen lexical items, so-called *lexical nodes* as analytic tools (Piecychna 2022b) will be used.[1] The analysed lexical nodes will then be extrapolated to other potential sensorial experiences which could be triggered by the particular lexemes. I am also inspired by Nikolajeva's (2017: 66) approach to the reader's experience with the text and posit that embodied aesthetic experiences do not 'necessarily happen to every actual reader; however, text affordances create a favourable condition' for a particular embodied aesthetic encounter to occur in the reading process.

Additionally, for the purposes of the intended analysis, Marković's (2012) approach to aesthetic information processing, with the following three components as fundamental aspects of the process – *attentive*, *cognitive* and *affective* – will be employed. According to the author,

> A person is fascinated with a particular object, whereas the surrounding environment is shadowed, self-awareness is reduced, and the sense of time is distorted. Amplified arousal and attention provide the additional energy which is needed for the effective appraisal of symbolism and compositional regularities in 'virtual' aesthetic realities. Finally, during this process a person has a strong feeling of unity and the exceptional relationship with the object of aesthetic fascination and aesthetic appraisal.
>
> (Marković 2012: 12)

In harmony with Marković's (2012) concept of aesthetic experience, it is claimed in this chapter that the experiencer needs to be transcendentally moved from the sphere of the pragmatic to the sphere of the aesthetic by transferring that which is explicit towards that which is of a more implicit, or symbolic, nature. 'An encounter with a landscape narrative becomes an aesthetic experience when sensory input from the first stage of aesthetic information processing is enriched with symbolic dimensions of meaning, or, in other words, with higher semantic levels of the content' (Piecychna 2022b). Similarly to my previous publication (Piecychna 2022b), two main levels of analysis will be used: *sensorium* (or explicit sensory input), that is, 'the perceptual and cognitive appraisal of the object's basic properties' (Marković 2012: 6), and *non-sensorium* (or implicit, 'hidden', symbolic input), that is, 'the detection of more complex compositional regularities and the interpretation of more sophisticated narratives and hidden symbolism of the object's structure' (Marković 2012). This means that the bodily format enables the recipient of the text to comprehend and interpret the semantic content of the text which lies deeper and which indicates the interpreter's experience with the reading process as based on his/her cultural and social values.

The texts from which fragments to be analysed come are as follows[2]:

The source text	The target text no. 1	The target text no. 2
She opened her eyes and looked about her. They were on the crest of a hill. The sun had set some time since, but the landscape was still clear in the mellow afterlight. To the west of a dark church spire rose up against a marigold sky. **Below was a little valley and beyond, a long, gently-rising slope with snug farmsteads scattered along it.** (Montgomery 1865/2008: 29)	Znowu otworzyła oczy i rozejrzała się wokoło. Znajdowali się na szczycie wzgórza. **Słońce było już zaszło, ale krajobraz tonął jeszcze w świetle** (The sun had already set but the landscape was still drowning in the light). Na wschodzie ciemna dzwonnica kościółka zarysowywała się na złoto czerwonawym obłoku (reddish puff). **Niżej roztaczała się mała dolina, a po drugiej stronie długi, nieco stromy łańcuch wzgórz, usianych ładnymi dworkami** (Below a little valley was unfolding, and on the other side a long, slightly steep chain of hills studded with pretty manor houses). (Montgomery 1911: 35)	Otworzyła oczy i rozejrzała się wokół. Znajdowali się na wierzchołku wzgórza, a chociaż słońce zaszło już jakiś czas temu, w zapadającym zmierzchu widać było wyraźnie całą okolicę (… and while the sun had already set some time ago, in the descending dusk the whole neighbourhood was clearly visible). Po zachodniej stronie, na tle wyzłoconego wciąż nieba (continuously gilded sky) rysowała się ciemna sylwetka kościelnej wieży. **Niżej, tuż za dolinką, łagodnie wznosiło się długie zbocze z malowniczo rozrzuconymi zadbanymi gospodarstwami** (Below, just beyond a little valley, a long slope was gently rising with well-kepts farms thrown here and there picturesquely). (Montgomery 2022: 35)

The fragments come from a description a scene where Anne Shirley sees her new home for the first time, which makes this experience full of strong emotions triggered in the protagonist (and, in accordance with the tenets of simulationism, in the reader of the description). Standing together with Matthew Cuthbert on the hill, the girl feels the atmosphere of the place.

Let us focus on the first sub-scene. In the source text, the protagonist, standing on the hill, has just seen the setting sun whose light has left a pleasantly smooth, heart-warming, almost dulcet afterglow in the sky. The use of the adjective *mellow*, as well as of the noun *afterlight*, seems to be employed on purpose in order to trigger in the reader positive associations with not only the landscape but also the protagonist. The target text from 1911, by Rozalia Bernsteinowa, reads as follows: 'Słońce było już zaszło, ale krajobraz tonął jeszcze w świetle' (The sun had already set but the landscape was still drowning in the light) (Montgomery 1911: 35). There are significant differences between the source and target messages. While in the original the author underlined that there has been some time since the sun set, indicating indirectly in this way the colour of the sky, which is very characteristic when the afterglow occurs, the Polish translator generalized and neutralized the scene by only slightly referring to what is usually happening after the sun has set. The other difference relates to the way of describing the landscape. While in the source text, Montgomery accentuated that the landscape, despite the sunset, was 'easy to see' (Macmillan Dictionary, online), Rozalia Bernsteinowa, by underlining the presence and importance of the light, which caused the landscape to submerge in it, transferred that which has been more explicit in Montgomery's description into the state of mere inexplicitness, or unclearness. In other words, while in Montgomery, what is in the foreground is the landscape itself and its explicit features, which help the reader to compose this particular spatial frame, in Bersteinowa this is the light which becomes dominant, persistent and causative.

Let us now concentrate on the latest Polish retranslation of Montgomery's novel, namely, that from 2022 by Anna Bańkowska. The fragment under focus in this translation reads as follows: 'a chociaż słońce zaszło już jakiś czas temu, w zapadającym zmierzchu widać było wyraźnie całą okolicę' (and while the sun had already set some time ago, in the descending dusk the whole neighbourhood was clearly visible) (Montgomery 2022: 35). While Bańkowska, similarly to Montgomery, underlined that some time passed between Anne Shirley and Matthew Cuthbert stood on 'the crest of the hill' and the sun had set, different aspects of the scene were foregrounded. Interestingly enough, Bańkowska did not even use the lexical item 'landscape' in her rendering. She used the more neutralized item *okolica* (neighbourhood), meaning in the first place 'an area stretching around a particular place' (Słownik Języka Polskiego, online), a definition different from that which is employed in order to refer to 'landscape', a lexeme meaning 'an area of land that is beautiful to look at or that has a particular type of appearance' (Macmillan Dictionary, online). These two definitions indicate that while *okolica* does not trigger associations with beauty directly, 'landscape' has within its semantic framework the notion of a beautiful area which brings about strong emotions, either positive, extremely pleasant sensations or those of a negative nature, for example, causing fear of the feeling of the sublime.

More importantly, however, in the latest Polish retranslation, Bańkowska translated 'mellow afterlight' as *zapadający zmierzch* (descending dusk), a solution which evokes a completely different image in the implied reader's mind than the source phrase 'mellow afterlight'. While 'mellow afterlight' triggers soft, warm, melodious, smooth, well-rounded sensations and experiences, the expression 'descending dusk' seems to trigger in the implied reader opposite associations and reactions to the depicted scene, that is, those pertaining to the darkness (or semidarkness), twilight, shadows and nightfall. Despite the fact that the two lexical items (afterlight and dusk) relate to the same phenomenon, they evoke completely different connotations: 'afterlight', due to the morphological process of compounding, already has within its structure the free morpheme 'light', which brings associations with the certain positive agency (because it means the light which one can see after the sunset), the lexical item 'dusk', however, indicates directly 'the time before night' (Macmillan Dictionary, online), thus having the potential of evoking in the implied recipient the lack of comfort or safety, lack of security or, generally speaking, negativity.

The simulation process is different in the source recipient and in the target recipients. The source recipient could simulate standing on the hill with Matthew Cuthbert and looking at the landscape touched by the delicate afterglow. The target recipient reading the first Polish translation could simulate standing on the hill with Matthew Cuthbert; however, in this case the implied reader could definitely notice more light surrounding the area. In the last case, the implied recipient could simulate standing on the hill in a much darker area than the previous two readers with a lower level of emotionality as associated with what s/he could have in their vision. Also, as based on the above considerations, one could state that the recipient, following the lines of aesthetic information processing, moves from the pragmatic (what is seen) towards the symbolic (what is perceived implicitly and revealed from the surface). The source text and the first target text are similar in their transference from the pragmatic to the symbolic (seeing the particular landscape which evokes positivity, beauty and optimism); however, the second target text is different from the previous two in terms of both the pragmatic (what is seen is not exactly the landscape but the loosely interpreted neighbourhood) and the symbolic (the transference into hidden symbolism implies certain agency of negativity and darkness).

The next frame reads as follows: 'Below was a little valley and beyond, a long, gently-rising slope with snug farmsteads scattered along it' (Montgomery 2008: 29). Here Anne Shirley, after having noticed the spire and the marigold sky, looks at the valley situated below the church, and then her eyes are set on a gentle surface going upwards, where cosy buildings located within the areas around farms are thrown here and there on the slope. Again, very positive emotions and associations are evoked by means of the following lexical items: 'little', 'gently-rising' and 'snug'.

The first Polish translation, from 1911, goes as follows: *Niżej roztaczała się mała dolina, a po drugiej stronie długi, nieco stromy łańcuch wzgórz, usianych ładnymi dworkami* (Below a little valley was unfolding, and on the other side a long, slightly steep chain of hills studded with pretty manor houses, Montgomery 1911: 35). Bernsteinowa, by using the verbal phrase *roztaczała się* (was unfolding), makes the implied reader see the vastness, expansiveness and dominance of the valley, which inadvertently creates dissonance between the two texts as in the original the valley is little and is used in the description only as the background of the farmsteads. Also, the dissonance is strengthened by the fact that in the source text a 'little valley' is positioned further from the beginning of the sentence as compared with the position of the counterpart expression in the Polish rendering, thus making the Polish phrase more prominent in the reading process.

The vastness in the Polish rendering is also manifested on yet another level. While Montgomery used the phrase a 'long, gently-rising slope', Bernsteinowa translated it as *nieco stromy łańcuch wzgórz* (a slightly steep chain of hills). The pluralia tantum used by the Polish translator makes the whole area seem much more expansive than in the source text. Also, the Polish item *łańcuch* (chain) triggers the implied reader associations pertaining to a relatively large number of slopes (while in the original, only one slope is prominent in the text). Besides, Montgomery described the slope as 'gently-rising', indicating that the slope is going up rather mildly. Bernsteinowa, however, accentuated that the chain of hills was falling and rising at a rather sharp angle. The implied reader then simulates noticing a number of rather perpendicular hills within the vastness of a little valley, while in the original, the implied reader could simulate observing one gently-rising slope within a rather small area with clearly marked boundaries.

Finally, Montgomery described the houses as 'snug farmsteads', a phrase which could be further retranslated into English as small, warm, cosy and comfortable main buildings located within the farm area. Interestingly enough, in Bernsteinowa's rendering the 'sung farmsteads' were translated as *ładne dworki* (pretty manor houses). While 'farmsteads' trigger clear and direct associations with the agriculture, *dworek* (manor house) is a strong culture-specific item for Poles for whom the buildings are traditionally the epitome of Polish nobility, aristocracy and higher classes. These were centres of cultivating symbols and traditions as ingrained within Polish culture and fighting for the independence of Poland, in particular within the time period from the end of the eighteenth century till 1918 (for more on this aspect of Bernsteinowa's translational solution, see Piecychna 2022a).

There is also another significant difference between the source text and the translation in terms of the distribution of the building. While in the source texts the farmsteads are thrown randomly below the valley, in the translation they are studded, meaning that there are many such manor houses, that the slope is almost augmented with them, a scene indicating a rather large number of the houses. This makes the aesthetic experience utterly different in the implied reader in the source text and in the target text. More importantly, however, the landscapes with farmsteads and that with manor houses evoke utterly different visions. The transference from the pragmatic towards the symbolic is here different in the source and target texts. In Montgomery's novel, the symbolic nature of the sub-scene implies the agricultural area, as well as the simplicity of life and poverty (the province itself belonged then to the poorest in Canada). Bernsteinowa's rendering, however, indicates completely different symbols, namely, tradition, nobility, beautiful architecture, higher classes, eloquence and wealth. Without too gross a generalization, then, one could argue that the two sub-scenes, in particular their final components with the type of dwelling, lie on two opposite sides of the same spectrum.

Bańkowska's translation of the analysed sub-scene reads as follows: *Niżej, tuż za dolinką, łagodnie wznosiło się długie zbocze z malowniczo rozrzuconymi zadbanymi gospodarstwami* (Below, just beyond a little valley, a long slope was gently rising with well-kepts farms thrown here and there picturesquely, Montgomery 2022: 35). On the one hand, Bańkowska's way of depicting the landscape, despite certain discrepancies in localizing the farm buildings, is quite close to that of Montgomery as the translator accentuated the length and gentleness of the slope. On the other hand, though, while Montgomery intended for the reader to visualize the cosiness of the buildings forming farms, Bańkowska underscored that the farms were well-kept, information which is not present in the source message, and not 'snug', as well as focusing on how picturesquely (rather than randomly) the buildings were thrown along the slope, thus creating an utterly different landscape scene for the implied reader to see.

In other words, while in the original, the intended recipient could simulate observing nice farmsteads, triggering warm and comfortable sensations and being situated randomly along the gently rising slope, in the translation the target implied reader could rather simulate looking at the tip-top condition of the farm buildings and concentrating on the picturesqueness of the simulated aesthetic experience. To put it differently, the source text retains its more pragmatic nature and simplicity of existence of the protagonists living within the area as described in this particular sub-scene. However, the target text transforms the simple character of the area into a more idyllic, almost blissful place. It should be underlined, though, that the translation has here created a certain contrast by juxtaposing the orderly character of the farms with the picturesqueness of the area surrounding the buildings.

Conclusions

Authors (including translators) and readers are active culture-specific meaning generators. They always perform an active role in reshaping and recontouring the scope and specificity of the spatial and (geo)cultural structures that surround them daily. The reader intervenes in the process of creating meaning-specific entities as they relate to the scope and nature of a given description of the landscape, which is determined by the historical times in which the landscape is perceived and the cultural context within which a particular segment of text is created.

By employing the paradigm of embodied aesthetics in the translational discourse in a systematic way, it might be possible to investigate ways in which landscape has been processed by a given society. This type of analysis which has been put forward in this chapter – where linguistic and cultural analysis is intertwined with a concern with the representation of a given fragment of described reality in the reader's mind – may show potential for the analysis of literary translation within not only translation studies as a whole but also within translational stylistics in particular. At the risk of simplification, one can state that the third decade of the twenty-first century could well become the time in which the cognitive literary turn within translation studies is more widely recognized. It is only to be hoped that this preoccupation with both the aesthetic and the embodied in literary translational analyses will pervade future work within the field of translational stylistics.

This chapter has argued that stylistic analyses within translation studies must be deepened by considering a wide array of factors pertaining to the functioning of the human mind and to aesthetic frameworks associated with the reader's involvement with a (translated) literary work. Such a focus on cognitive (including neural) components enables one to conduct more in-depth translational analyses of literary renderings. Besides, a genuine form of interdisciplinary cooperation between translational stylistics and the paradigm of an (aesthetic) embodiment should be taken on board in order to deploy the results in an adequate and satisfactory way, all the more so that the embodiment paradigm is still being explored by representatives of cognitive studies, and the conclusions that they reach should not be applied to humanistic disciplines uncritically.

The main purpose of this chapter was to document the ways in which the same description of the landscape is represented in two Polish translations of L. M. Montgomery's *Anne of Green Gables*: the earliest one (from 1911) and the latest one (from 2022). From the analysis above, it transpires that landscape is represented differently, not only cross-culturally (the source text – the target text dynamics), but also cross-historically (the *nth* target text – the *nth* target text dynamics). The differences between the source text and the target texts, as well as between the two Polish translations themselves, could be said to be caused by historical, cognitive, cultural and social contexts within which the author of the source text and the translators were embedded. While this chapter offers new ways of analysing literary renderings by means of the embodied approach to translation, it is, though, not free from methodological flaws as its focus was mainly placed on the implied reader, a generalized experiencer whose encounter with the text can be extrapolated to the broader receiving audience.

Besides, the bodily format that here was treated as an analytical component cannot serve as a tool which could yield any objectivized results. Despite these disadvantages an important conclusion can be reached, namely that cognitive literary translation studies should be pursued by translational statisticians, which can be successfully initiated by mixing the linguistic with the cognitive.

Notes

1   '*Lexical nodes* should be understood as trigger-lexical items, including phrases, which are particularly prone to evoke certain reactions in the reader's sensorimotor and perceptual systems during the act of reading. The quality of being particularly

prone to trigger embodied reactions means that selected lexical items are associated with the functioning of the sensory apparatus of the human being. And thus such words, concrete rather than abstract, should include those items that could be seen, heard, touched, smelled or tasted by the individual. It is also assumed, in line with Minsky's theory (1988), that the lexical items activate frames of semantic knowledge pertaining to the concepts they represent' (Piecychna 2022b). The lexical nodes which will be chosen for analysis form certain scenes as well as sub-scenes which together create a concrete landscape frame to be experienced by the protagonist, the translator and the implied reader.

2   Due to space constraints only those fragments of the texts will be analysed in detail which are marked in bold.

References

Baldo, M. (2013), 'Landscapes of return: Italian-Canadian writing published in Italian by Cosmo Iannone Editore', *Translation Studies*, 6 (2): 199–216.

Bergen, B. K. (2012), *Louder than Words: The New Science of How the Mind Makes Meaning*, New York: Basic Books.

Bergen, B. K. and K. Wheeler (2010), 'Grammatical aspects and mental simulation', *Brain and Language*, 112: 150–8.

Boase-Beier, J. (2006), *Stylistic Approaches to Translation*, Manchester: St Jerome Publishing.

Boase-Beier, J. (2014), 'Stylistics and translation', in M. Burke (ed.), *The Routledge Handbook of Stylistics*, 393–407. London: Routledge.

Bruhn, M. J. and D. R. Wehrs, eds (2013), *Cognition, Literature, and History*, London/New York: Routledge.

Buffery, H. (2013), 'Negotiating the translation zone: Invisible borders and other landscapes on the contemporary "heteroglossic" stage', *Translation Studies*, 6 (2): 150–65.

Burke, M. and E. T. Troscianko, eds (2016), *Cognitive Literary Science. Dialogues Between Literature and Cognition*, Oxford: Oxford University Press.

Crane, M. T. (2001), *Shakesperare's Brain. Reading with Cognitive Theory*, Princeton: Princeton University Press.

Cronin, M. (1996), *Translating Ireland. Translation, Languages, Culture*, Cork: Cork University Press.

Cronin, M. (2000), *Across the Lines: Travel, Language, Translation*, Cork: Cork University Press.

Cronin, M. (2003), *Translation and Globalization*, London: Routledge.

Cronin, M. and S. Simon (2014), 'Introduction: The city as a translation zone', *Translation Studies*, 7 (2): 119–32.

Easterlin, N. (2012), *A Biocultural Approach to Literary Theory and Interpretation*, Baltimore: The Johns Hopkins University Press.

Farahzad, F. and S. Ehteshami (2019), 'Spatial Territories in Translation Studies', *Translation Studies Quarterly*, 16 (63): 71–87. Retrieved from https://journal.translationstudies.ir/ts/article/view/635 (7 February 2022).

Gallese, V. (2012), 'Embodied simulation theory and intersubjectivity', *Reti, Saperi, Linguaggi*, 4 (2): 57–64.

Gallese, V. (2017), 'Visions of the body. Embodied simulation and aesthetic experience', *Aisthesis. Pratiche, linguaggi e saperi dell'estetico*, 10 (1): 41–50.

Gallese, V. (2019), 'Embodied simulation. Its bearing on aesthetic experience and the dialogue between neuroscience and the humanities', *Gestalt Theory*, 41 (2): 113–28.

Hephzibah, I. (2013), 'Islam translated: Literature, conversion, and the Arabic cosmopolis of South and Southeast Asia', *Translation Studies*, 6 (2): 249–52. https://www.macmillandictionary.com/ (accessed on 25 February 2022) https://sjp.pwn.pl/ (accessed on 25 February 2022).

Italiano, F. (2016), *Translation and Geography*, London: Routledge.

Jaén, I. and J. J. Simon (2013), *Cognitive Literary Studies: Current Themes and New Directions*, Austin: University of Texas Press.

Kershaw, A. and G. Saldanha (2013), 'Introduction: Global landscapes of translation', *Translation Studies*, 6 (2): 135–49.

Kok, K. and A. Cienki (2017), 'Taking simulation semantics out of the laboratory: towards an interactive and multimodal reappraisal of embodied language comprehension', *Language and Cognition*, 9 (1): 1–23.

Marinetti, C. and M. Rose (2013), 'Process, practice and landscapes of reception: An ethnographic study of theatre translation', *Translation Studies*, 6 (2): 166–82.

Marković, S. (2012), 'Components of aesthetic experience: Aesthetic fascination, aesthetic appraisal, and aesthetic emotion', *i-PERCEPTION*, 3: 1–17.

Minsky, M. (1988), *The Society of Mind*, New York: Touchstone.

Montgomery, L. M. ([1865] 2008), *Anne of Green Gables and Anne of Avonlea*, Ware: Wordsworth Classics.

Montgomery, A. (*sic*) (1911), *Ania z Zielonego Wzgórza*, trans. R. Bernsztajnowa, Warszawa: M. Arcta.

Montgomery, L. M. (2022), *Anne z Zielonych Szczytów*, trans. A. Bańkowska, Warszawa: Marginesy.

Nikolajeva, M. (2014), *Reading for Learning: Cognitive Approaches to Children's Literature*, Amsterdam: John Benjamins.

Nikolajeva, M. (2017), 'Haven't you ever felt like there has to be more? Identity, space and embodied cognition in young adult fiction', *Encyclopaideia. Journal of Phenomenology and Education*, XXI (49): 65–80.

Piecychna, B. (2022a), 'Sophisticating the image of Avonlea in the earliest Polish translation of *Anne of Green Gables* by Lucy Maud Montgomery', *Perspectives: Studies in Translation Theory and Practice*, 30 (2): 209–23.

Piecychna, B. (2022b), 'Translation studies meets embodied aesthetics: On the narratives of landscape and national identity in *Anne of Green Gables* and its Earliest Polish Rendering', *SKASE. Journal of Translation and Interpretation*, 15 (1): 58–74.

Richardson, A. (2010), *The Neural Sublime: Cognitive Theories and Romantic Texts*, Baltimore: The Johns Hopkins University Press.

Schulte, R. (2001), *The Geography of Translation and Interpreting*, Lewiston: The Edwin Mellen Press.

Semino, E. and J. Culpeper (2002), *Cognitive Stylistics: Language and Cognition in Text Analysis*, Amsterdam: John Benjamins.

Simon, S. (2008), 'Cities in translation: Some proposals on method', *Doletiana*, 2: 1–12.

Simon, S. (2012), *Cities in Translation*, Abingdon: Routledge.

Simon, S. (2015a), 'Returns on translation: Valuing Quebec culture', *Contemporary French and Francophone Studies*, 19 (5): 501–13.

Simon, S. (2015b), 'The translational life of cities', *The Massachusetts Review*, 56 (3): 404–15.

Simon, S. (2018), '*Saxa Loquuntur*: Prague's speaking stones', *Art in Translation*, 10 (1): 55–70.

Skomorokhova, S. (2013), 'The Belarusian literary landscape and translation "waves"', *Translation Studies*, 6 (2): 183–98.

Stecconi, U. (2010), 'What happens if we think that translating is a wave?', *Translation Studies*, 3 (1): 47–60.

Torresi, I. (2013), 'The polysystem and the postcolonial: The wondrous adventures of James Joyce and his Ulysses across book markets', *Translation Studies*, 6 (2): 217–31.

Vermeule, B. (2010), *Why Do We Care About Literary Characters?*, Baltimore: Johns Hopkins University Press.

Von Flotow, L. and R. M. Nischik, eds (2007), *Translating Canada. Charting the Institutions and Influences of Cultural Transfer: Canadian Writing in German/y*, Ottawa: Ottawa University Press.

Williams, D. and M. Marinkova (2015), 'Affective trans-scapes: Affect, translation, and landscape in Erín Moure's *The unmentioable*', *Contemporary Women's Writing*, 9 (1): 73–92.

Zunshine, L., ed (2010), *Introduction to Cognitive Cultural Studies*, Baltimore: The Johns Hopkins University Press.

Zunshine, L., ed (2015), *The Oxford Handbook of Cognitive Literary Studies*, Oxford: Oxford University Press.

Zwaan, R. and D. Pecher (2012), 'Revisiting mental simulation in language comprehension: six replication attempts', *PloS One*, 7 (12): e51382. DOI: 10.1371/journal.pone.0051382.

Part Two

To ECO: Nature, Culture, and Beyond

The second part of this collection covers the move to ECO. The five chapters included here deal with the ecology of language and translation, as well as provide examples of ekphrasis. In Chapter 6, Eva M. Gómez-Jiménez discusses the challenges of working with a large corpus of newspaper texts, which centres on economic inequality in the UK. Despite the fact that the chapter itself focuses on the methodological implications of working with such a corpus, Gómez-Jiménez makes it evident how the language examined represents the economic ecosystem. Although dealing with a literary text rather than newspapers, Michał Garcarz (Chapter 8) also touches upon world-creating characteristics of language when he analyses *Macbeth* translated into African American urban slang. He does so with a strong emphasis on what he calls eco-translation insights. He speaks in defence of intralingual translation, providing a detailed analysis of Tonia Lee's (2008) version of *Macbeth*.

In Chapter 7, Marcello Giovanelli looks closely at Wilfred Owen's poem 'Futility' (1920). Like Chapters 4 and 5 in Part I of this volume, this chapter also deals with descriptions of landscapes. However, Giovanelli focuses firmly on ECO, examining the interaction in the poem between the body of a dead soldier and the landscape around it. This chapter also highlights the specificity of landscapes in First World War literature.

The final two chapters here cover ekphrasis. Polina Gavin (Chapter 9) explores Margaret Atwood's *Cat's Eye* (1988). Her analysis aims at illustrating how the descriptions of imaginary art reflect childhood trauma of the main character, Elaine. Through her unreliable narrator who has dissociative amnesia, Gavin links ekphrasis to the explorations of PTSD. She does so via the descriptions of

Elaine's paintings which reflect her gradual regaining of her traumatic memories. Although Chapter 10 also takes an ekphrastic lens, Maria-Eirini Panagiotidou shows how Dante Rossetti's poetry offers landscape immersion and transcendence via its descriptions of art. Panagiotidou analyses in detail Rossetti's sonnet 'For a Venetian Pastoral by Giorgione (In the Louvre)' (1870), showing how the presence of deictic expressions, image schemas, conceptual metaphors, personification and lexical choices which emphasize bodily movement facilitate the reader's own sensorimotor involvement and contribute to their perception of ORIGO (deictic centre). The five chapters in Part II of this collection deal with a variety of fictional and non-fictional texts and approach the move to ECO differently, thus successfully showcasing how versatile an applied cognitive ecostylistics approach can be.

6

Methodological Implications of Building *The Corpus of News on Economic Inequality (1971–2020)*: Text-readable Data vs OCR Material

Eva M. Gómez-Jiménez

Introduction

Economic inequality, also referred to as *class* or *wealth inequality*, is a form of social imbalance that broadly mirrors the social variable of class, a defining feature of individuals and groups just like gender, sexual orientation, religion or country of origin. This type of inequality has, in general terms, three overlapping sources: *pay inequality* (deriving from payments from employment only), *income inequality* (deriving from the money obtained from employment, investments, savings, state benefits, pensions and rent) and *wealth inequality* (deriving from the total amount of assets of an individual or household) (The Equality Trust 2022).

Many reasons justify the importance of addressing economic inequality in academia. Studies in a variety of fields have demonstrated that it has an overall negative impact on society, not just contributing to health issues (Kondo et al. 2009; Offer et al. 2012; Burns et al. 2014), worse educational performance (Morrisson and Murtin 2013), higher levels of crime and corruption (Elgar et al. 2013) or more social insecurity (Corak 2016), but also driving us towards more deregulated economies, which ultimately makes society more prone to financial crises, inflation and lower labour productivity (Iacoviello 2008; van Treeck 2014), among others. Also, economic inequality is rising globally, as recently indicated in the World Inequality Report (Chancel et al. 2022), and the UK is not an exception, with sociologists, economists and historians having demonstrated that class inequities have increased in this country since the early 1970s (see e.g. Stiglitz 2012; Piketty 2014; Forsey 2017). In fact, the UK at present has one of the highest levels of inequality within the OECD, with just Lithuania, the United States, Turkey and Chile proving worse (OECD 2022).

All these signs of newly entrenched inequality are best understood within the framework of neoliberalism, as we live in a social order where economics has expanded to almost every other aspect of our lives, thereby very much determining decisions that affect education and the health or degree of peace in society (Sandel 2012). In this context, language is becoming more relevant than ever, as contemporary societies in developed countries enjoy an economic and (by extension) a social system that is very much 'knowledge-' or 'information-based' (Fairclough 2002). Following the ideas of social constructionism (e.g. Fowler 1991; Fairclough 1992, 1995; van Dijk 1988; 1998), language may play an important role in this situation since the language we use (and the particular choices we make, either consciously or unconsciously) determines how we understand the world and how we interact as a result of certain aspects, such as race, gender or, as in this case, social class. More specifically, the language used in mass media discourse may have influenced the attitudes and expectations of society by either facilitating the social changes that have driven the UK towards being a more unequal country today or, otherwise, obstructing those that might have helped to mitigate this situation. In other words, by means of dominant public discourses and the values that newspapers and other media articulate, the premise of this work is that British society today may have been encouraged to see and understand inequality in terms that are broadly congruent with a neoliberal mindset that is drawing us closer to a value system present in other also highly unequal societies in the world, such as in the United States, for example.

Bearing in mind the aforementioned, this chapter discusses a work in progress that forms part of a larger project concerning diachronic explorations of the discourse of economic inequality in the British press between 1971 and 2020 and that, broadly speaking, aims to foster research studies and scholarly discussion on this particular genre of discourse. To do so, this chapter is focused on the methodological implications of compiling a diachronic corpus of UK news on economic inequality during the period under consideration, although it also presents some ideas on its further exploration within the context of corpus-assisted discourse studies (CADS). The following sections deal with the previous approaches employed to examine the discourse of economic inequality and economic-related issues in Britain, the usefulness of CADS in a project of this nature, the data to gather and the process involved in compiling and inspecting *The Corpus of News on Economic Inequality (1971–2020)*. Finally, some discussion and closing remarks are provided.

The discourse of economic inequality in the British context

Although forms of economic inequality have led to growing interest in academia since the late 2000s (Machin and Richardson 2008), studies from a critical discourse perspective are in relatively short supply, especially if we compare them to approaches concerning other societal issues such as gender, race, sexual orientation or religion. This increased interest is evidenced by the few studies to date that have inspected economically related issues through their discoursal representations in the British context, some of which are collected in special issues in the journals *Discourse & Society* (Fairclough 2002) and *Critical Discourse Studies* (Machin and Richardson 2008; Silke and Rieder 2019). Accordingly, there are numerous studies on the uses and users of the benefits systems, which, as the results suggest, are normally portrayed in a negative way in both the mass (see e.g. Roberts 2017; Incelli 2021) and social media (see e.g. Baker and McEnery 2015; Paterson, Coffey-Glover and Peplow 2016; Paterson, Peplow and Grainger 2017). In a similar vein, approaches to the discourse of poverty suggest that the poor are portrayed as a security threat (Lorenzo-Dus and Marsh 2012), in a general trend that normally conveys them negatively in the media (see especially Meinhoff and Richardson 1994). Supplementing these, other studies have demonstrated that, conversely, there is a public defence of those more affluent in UK society, which includes the royal family, celebrities, big corporations and the rich (see e.g. Graham and O'Rourke 2019; Rieder and Theine 2019). Further work have also acknowledged a recent denial of class struggle in UK society and the idea that class has disappeared from the political agenda in this country (see e.g. Grisold and Silke 2019), together with a recent display of austerity measures that are conveyed as inevitable and morally right (see e.g. Fairclough 2016).

Very few approaches, though, have looked particularly at how the representation of forms of class inequality in the UK has changed since the 1970s (see especially Baker forth., Jeffries and Walker 2019). In this way, this chapter, as acknowledged above, is part of a larger body of corpus-assisted work (see Toolan 2016; Gómez-Jiménez 2018; Toolan 2018; Gómez-Jiménez and Toolan 2020; Gómez-Jiménez and Bartley, 2023) that intends to explore the representation of forms of class inequality in the last fifty years. Findings in this area so far have shown that the concept of class had mostly disappeared in 2013 T.V. reviews in the *Daily Mail* (Toolan 2016), and that discussions about maternity leave benefits became monetized in the *Times* and *Daily Mail* in the late 1990s (Gómez-Jiménez 2018).

From a broader perspective, Toolan (2018) has identified a number of relevant linguistic patterns that implicitly changed the representation of this form of inequality in recent decades in the *Times* and *Daily Mail*. Similarly, Gómez-Jiménez and Toolan (2020), in their co-edited volume, have invited contributors to inspect propaganda, newspaper, political and television discourse and noted that all of these helped in making sharply increased wealth inequality seem perfectly normal. More recently, Gómez-Jiménez and Bartley (2023) have also analysed the representation of homeless people and homelessness in *The Guardian* and *Daily Mail* from 2000 to 2018, with results suggesting that, within an overall negative representation, the focus in more recent years has shifted towards implying that homelessness occurs inevitably, and therefore is not an issue that the government or society needs to address. In light of the research studies to date, then, this chapter aims to offer a further contribution to the literature, the focus here being on the methodological implications of compiling a diachronic corpus of this type and especially in view of how this form of inequality has steadily increased in more recent years in the UK.

Corpus-assisted discourse studies

Critical Discourse Analysis (CDA) is at the core of this chapter, although I also draw on Corpus Linguistics (CL) to complement the former. In this respect, and following a recent trend in recent years, CADS has been defined as 'that set of studies into the form and/or function of language as communicative discourse which incorporate the uses of computerized corpora in their analyses' (Partington, Duguid and Taylor 2013: 10). The origin of CADS is often traced back to Mautner (1995) and Stubbs (1996), though other antecedents include Caldas-Coulthard (1993), Louw (1993), Stubbs and Gerbig (1993), among others (see Mautner 1995: 2).

The main features of CADS, and therefore the ones that make this framework useful in work of this nature, are best summarized by Partington, Duguid and Taylor (2013), who highlight (i) the combination of quantitative and qualitative approaches to discourse, whereby statistical and computerized overviews of large amounts of a particular genre of discourse are combined with the close analysis of particular samples of such discourse; (ii) its aim to try to analyse and critique the discourse type in question as much as possible, which means approaching the corpus in a wide variety of ways; (iii) its use of, normally, specialized corpora that are built specifically according to the research questions or hypotheses that researchers aim to address; and (iv) its comparative nature, based on the notion that you can only approach and evaluate a corpus (or discourse) if you compare it to another akin (2013: 11–14). This leads us to employ many different and possible methods when inspecting a corpus, where the choice of one or another will depend on the research questions and/or the hypotheses that ground our research.

As with CDA, CADS' overarching aim is to uncover 'non-obvious' meanings, or meanings which are not perceived at first sight, in the discourse under scrutiny (Partington 2010: 88). These non-obvious meanings in discourse occur because, in speaking or writing, we make semi-automatic choices in the language we use, those of which come to the forefront when applying transitivity, modality or vocabulary analysis (among many others), these analyses being further facilitated when we adopt a combination of quantitative (statistical) and qualitative (in-depth) tools. In this sense, CADS draws together what can be learned from corpus analysis with other sources of information on a particular topic, whether they be linguistic or socio-cultural (Partington 2010: 90). Software tools are incorporated here because they have a potential 'in helping to unravel how particular discourses, rooted in particular socio-cultural contexts, construct reality, social identities and social relations' (Fairclough 1992: 64). It is important to highlight at this point that, notwithstanding that the use of corpus tools complements the methods normally used in CDA, these should not replace them, as 'qualitative and quantitative techniques need to be combined, not played one against each other' (Mautner 1995: 2).

Although there is not a single method to undertake a corpus-assisted discourse study, it is possible to identify a few standard steps in the process (see Mautner 1995) that can be succinctly summarized as (i) building (or gathering), (ii) annotating and (iii) analysing the corpus. At the analysis stage, countless options exist for the analyst, who can combine concordancer tools with additional software and, thus, have each tool constantly feed into the other. This also means that the hypotheses and/or research questions that instigated the study will further impact any methodological decisions, as well as have theoretical implications for the research in question. All that said, this chapter discusses the first step in this aforementioned process, that is, the building or gathering of *The Corpus of News on Economic Inequality (1971–2020)*, commonly abbreviated as *DINEQ* for *Discourse* and *INEQuality*.

Data

The Corpus of News on Economic Inequality (1971–2020) will consist of British national newspaper articles about economic inequality from 1971 to 2020. The corpus will contain newspaper data, rather than other types of media data, following Toolan's (2018) triple claim (i) that the most relevant newspapers are digital and online, as well as printed, (ii) that the best and most persuasive journalists still write for the national newspapers, and (iii) that people, despite using social media to stay informed, still turn to newspapers, T.V. news and radio channels for in-depth views on social affairs. There is also the fact that the nature of this project itself, which aims to look at discourse in the past fifty years, demands this kind of media be inspected, as there was no social media back in the 1970s to compare present-day discourse to. Moreover, news brands in the UK are still reaching 88 per cent of the adult population through either printed or digital means, according to recent estimates (PAMCo 2022).

The *DINEQ* corpus will collect samples from national newspapers in the belief that these are the ones that have, if so, shaped people's attitudes and expectations regarding economic inequality to a greater extent. Finally, the corpus will contain data from newspapers, regardless of political orientation, in order to provide a representative sample of the British press over the period under consideration. Furthermore, this ensures a wide variety of material to facilitate further discussion and research studies on the topic, assuming some scholars might be more interested in comparing the discourse across the newspapers' political alignment than in looking for possible discoursal changes across time (despite this not being the focus of this research).

Although there exists a wide variety of databases containing newspaper material, there is not, to my knowledge, one that provides access to the whole period in question in this project, that is, from 1971 to 2020. For this reason, the compilation of *DINEQ* will include the use of a full-text database in combination with the newspaper historical archives. The data through these will follow a similar process and the same criteria will be applied to ensure that a high-quality and balanced corpus is retrieved; nonetheless, it will inevitably imply adapting the process in line with the peculiarities of each archive. As soon as the data have

Table 6.1 Overall process for the compilation of the DINEQ corpus

Step	Main tasks	Resulting material
1	Gathering of text-readable data	Dineq_online
2	Gathering of OCR data	Dineq_historical
3	Assembly of Dineq_online and Dineq_historical + final arrangements	DINEQ corpus

been assembled and the corpus is ready, it will be made available for scholars interested in exploring the discourses surrounding economic inequality (see Table 6.1).

The gathering of text-readable data will be done through *Nexis Uni* (2022), which is an online database that, among others, provides access to a large international news archive. *Nexis Uni* offers a convenient search engine for a project like this since it gives large amounts of data in a manageable way. Beyond that, there are sound reasons to use this database in particular. Firstly, it provides comprehensive access to the most-read national newspapers in the UK according to the Audit Bureau of Circulations (2022), including *The Guardian, The Independent, The Daily Mail, Metro* or *The Times*, among others (see Table 6.2). It is worthwhile remarking that all these sources represent both different political orientations despite their possible ideological fluctuations over time (from very left-wing to very right-wing), as well as different target audiences (tabloids, midmarket and quality press). Secondly, *Nexis Uni* provides digitalized, text-readable data that allow for a fairly automatized gathering process and the direct retrieval of results as .txt files, thereby ensuring a relatively simple process at this stage.

Nexis Uni is not free of limitations, though. Its main pitfall is the fact that the access it provides to each of Britain's newspapers is inconsistent: while data from *The Guardian* stem back to July 1984, the ones from *Metro* stem back to as recent as December 2007, for instance. In any case, this database does not include any data from UK national newspapers that were printed before July 1984, which means that the corpus in this project will inevitably require the use of OCR material.

In order to complement the text-readable data available on *Nexis Uni*, I will turn to different newspaper archives that offer OCR material from the earliest decades. In particular, the data will be obtained through *Gale Historical Newspapers* (2022) (which gives access to the *Daily Mail, Independent, Mirror*

Table 6.2 Access to UK national newspapers through *Nexis Uni*

Newspaper	From	Through
Metro	5 December 2007	Current
The Sun	1 January 1996	Current
The Daily Mail and *Mail on Sunday*	1 January 1992	Current
The Evening Standard	2 January 1992	Current
Daily Mirror and *Sunday Mirror*	29 May 1995	Current
The Times and *Sunday Times*	1 July 1985	Current
The Daily Telegraph and *Sunday Telegraph*	30 October 2000	Current
Daily Star and *Daily Star Sunday*	1 December 2000	Current
The Express and *Sunday Express*	2 October 1999	Current
The Independent and *Independent on Sunday*	19 September 1988	Current
The Guardian and *The Observer*	14 July 1984	Current
Daily Record and *Sunday Mail*	1 January 1994	Current

and *Telegraph* archives), *Newspapers.com* (2022) (which gives access to the *Guardian*, *Observer* and *Evening Standard*) and *UK Press online* (2022) (which gives access to the *Express* and *Daily Star*).

Method

Corpus compilation

The first step in compiling *DINEQ* consists in selecting the list of keywords for the data search in both *Nexis Uni* and the historical archives mentioned above. Parting from the definition provided by The Equality Trust (2022), the selected initial terms include *economic inequality, class inequality, social inequality, pay inequality, income inequality* and *wealth inequality*. These, conversely, can be transformed into the following core query: *econom* OR class OR wealth OR pay OR income OR social AND ineq**. This guarantees that the data search will find news items containing mainly *inequality* and *inequity*, but also related terms such as *inequability, inequable, inequal, inequalitarian, inequation* or *inequitable*, although only if they appear in conjunction with the former economic-related terms. Additionally, this core query is less restrictive than using, for instance, *economic inequality* directly, which means that the search will encounter a higher number of results but they will remain relevant to the research objectives in this project.

Once the core query is ready, I will apply it to the search engine in *Nexis Uni* using the advanced news search tool. Beyond the query itself, the search will be restricted through the following parameters: the query will be applied to all fields in the text, and the date field search option will be open to any result between 1 January 1971 and 31 December 2020. The search will be done for every newspaper source independently, as per the ones included in Table 6.2. The results will then be sorted by date (newest to oldest) and manually inspected to avoid duplicates and results unrelated to the British context. They will be downloaded as .docx files and labelled individually according to the newspaper and specific date of publication: for example, 20201223_Times. Finally, they will be converted into .txt files (UTF-8 coded), so that they can be subsequently uploaded to different corpus tools such as *Antconc* (Anthony 2022), *UAM Corpus Tool* (O'Donnell 2016) or *Sketch Engine* (Kilgarriff et al. 2014), and cleaned to remove unnecessary information, such as the copyright line, the page number, the length info, the language or the publication types.

Once all of the news available through *Nexis Uni* have been gathered, the next step will consist in applying the core query to the search engine in the historical archives that provide access to the missing news items in the former, mostly corresponding to the 1970s, 1980s and 1990s. The main difference between the data from earlier and more recent decades, and probably the most problematic issue here, is the fact that items in the historical archives are provided as OCR material, which means that most of them have been scanned from microfilm into .pdf or similar graphic formats, having been indexed in searchable text databases through optical character recognition (OCR) technology. For this reason, the data will be downloaded initially as .pdf files, where possible, and then converted into .txt files through *Abbyy FineReader* (2022), an OCR application that allows the conversion of image documents into editable file formats (see Figure 6.1). After gathering the older news items through the historical archives, I will conclude by assembling all the collected data together and, if necessary, mass-edit it through source code editors such as *Notepad++* (Ho 2022).

Corpus inspection

The second step to be taken after compiling *DINEQ* will consist of inspecting the corpus itself. The initial idea is to examine news items across time (e.g. by comparing news items during the economic recessions in the UK, that is, the mid-1970s, the

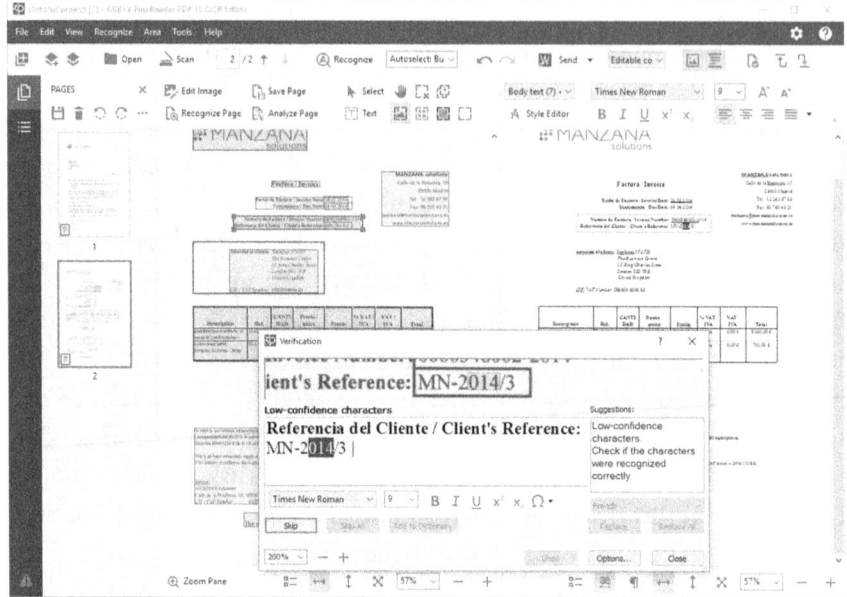

Figure 6.1 *Abbyy FineReader* interface (2022).

early 1980s, the early 1990s and the late 2000s), and look for possible discoursal changes. As outlined above, there is not a single, standard process of analysis within CADS, with authors suggesting the methodology is reflexive and abductive; as such, DINEQ will undergo two main stages: a quantitative search for statistically significant keywords followed by a further, more qualitative analysis.

In order to obtain statistically significant keywords in the newspapers discussed, those which will serve to look for potential areas of interest in the corpus, I will first use *Antconc keyword list*, which is a function available through *Antconc* (Anthony 2022) that shows unusually frequent (or infrequent) words in the periods investigated. Following this, I will calculate the log ratio values (Hardie 2014; Gabrielatos 2018) of the most significant keywords. Log ratio is an 'effect size' statistical measure that represents how big the difference between two corpora is for a particular keyword; thus, a word whose relative frequency in the two corpora is exactly the same will receive a log ratio value of 0; if it is two times more common in one corpus than in the other, then it will receive a value of 1; if it is four times more frequent in one corpus, it will receive a value of 2, and so on (Hardie 2014).

For the more qualitative analysis of the most significant keywords, I will follow the approach adopted by Baker (2006), who proposes the following: firstly, taking fifty randomly selected samples and analysing them through the concordance lines (e.g. through *Antconc*); then, repeating the same process with another fifty samples, testing the patterns found previously and (maybe) find new ones; finally, repeating this process until no new patterns emerge. This particular method implies a kind of data-driven analysis, which means that the linguistic aspects subject to analysis will be determined according to initial and subsequent results. Examples include, but are not limited to, thematic analysis (van Dijk 1995), transitivity patterns (Halliday and Matthiessen 2014), collocational analysis (Baker, Gabrielatos and McEnery 2013; Baker 2014), textual opposition (Jeffries 2010; Davies 2013) or metaphor (Hart 2008; Mussolf 2012), to name a few.

Discussion and conclusion

This chapter has discussed the compilation and inspection of *DINEQ, The Corpus of News on Economic Inequality (1971–2020)*, as part of a larger project that aims at fostering discussion on the discourse of economic inequality. This topic has traditionally received less attention within critical discourse studies by comparison to other social variables such as genre, religion or sexual orientation. When ready, *DINEQ* will constitute a diachronic corpus of printed and online British newspaper material on the topic of economic inequality from the years 1971–2020. It will include mainly news, although also other subgenres such as opinion articles or letters to the editor. It will be organized into subcorpora corresponding to the newspaper and year in question since, discursively speaking, the main focus in the project is on the possible changes that have occurred in the discourse of the British press across time. Although, in principle, the corpus will contain no metadata, the obtention of funding and support in the near future will also mean that this information can be included during the final stages of compiling the corpus. Lastly, with the focus here on the methodological implications of building this corpus, using a combination of online databases with historical archives for the period under consideration, with all the difficulties this entails, is unavoidable.

As outlined here, there are two fundamental issues that need to be resolved before undertaking this work in the near future. Firstly, the main issue with building a corpus of this nature is the amount of time and effort required for this endeavour since it covers a lengthy timespan (i.e. 1971–2020), thereby giving an increased number of results in online databases. When comparing the search results for *The Guardian*, for example, *Nexis Uni* gives forty-six items for the year 1984, while it offers 2,188 for the year 2020. This, roughly speaking, corresponds to almost fifty times more results in the more recent period and, therefore, reflects the need to find other scholars interested in collaborating on this project before it is launched. Secondly, the initial query presented here, which departs from the official definition provided on the concept of economic inequality by The Equality Trust (2022), is taken as a tentative one, and is open to including additional terms (see e.g. Gabrielatos 2007; Perez-Paredes 2017) such as synonyms or semantically related collocates, as well as other terms within economic theories (e.g. *welfare, capital*) or even deriving from one's common sense (e.g. *poor, aristocracy, homeless, ruling class*). Although I will try to adhere to the original definition as much as possible, I am open to expanding the query terms before corpus compilation commences.

Broadly speaking and bearing in mind the fact that this endeavour forms part of a larger research project, there are a number of steps outstanding before this work can be completed. Firstly, the project will assemble a team, formed primarily of scholars in the fields of CDA and CL, but also ideally comprising researchers in other related fields such as economics and the media. Secondly, the project will be presented to a number of research funding programmes at a regional (e.g. Andalusian Plan for Research Development and Innovation), national (e.g. the Spanish Research Agency National Plan) and European levels (e.g. ERC grants). If funding is obtained, we will officially launch the project and, as a point of departure, proceed to polish the data search criteria and the methodology discussed in detail in this chapter.

References

Abbyy FineReader (2022), [Computer Software]. Available online: https://pdf.abbyy.com/ (accessed 24 November 2022).

Anthony, L. (2022), *AntConc* [Computer Software], Tokyo, Japan: Waseda University. Available online: https://www.laurenceanthony.net/software (accessed 24 November 2022).

Audit Bureau of Circulations (2022). Available online: https://www.abc.org.uk/ (accessed 22 November 2022).

Baker, P. (forthcoming), 'Making the needy look greedy: Using corpus methods to examine The Sun's discourse around benefits', in J. Rahilly and V. Vander (eds), *Crossing Boundaries: Interdisciplinarity in Language Studies*, Amsterdam: John Benjamins.

Baker, P. (2006), *Using Corpora in Discourse Analysis*, London: Continuum.

Baker, P. (2014), *Using Corpora to Analyse Gender*, London: Bloomsbury.
Baker, P., C. Gabrielatos and T. McEnery (2013), *Discourse Analysis and Media Attitudes: The Representation of Islam in the British Press*, Cambridge: Cambridge UP.
Baker, P. and T. McEnery (2015), 'Who benefits when discourse gets democratised? Analysing a twitter corpus around the British Benefits Street debate', in P. Baker and T. McEnery (eds), *Corpora and Discourse Studies*, 244–65. London: Palgrave.
Burns, J. K., A. Tomita and A. M. Kapadia (2014), 'Income inequality and schizophrenia: Increased schizophrenia incidence in countries with high levels of income inequality', *International Journal of Social Psychiatry*, 60 (2): 185–96.
Caldas-Coulthard, C. (1993), 'From discourse analysis to critical discourse analysis: The differential re-presentation of women and men speaking in written news', in G. Fox, M. Hoey and J. M. Sinclair (eds), *Techniques of Description: Spoken and Written Discourse*, London: Routledge.
Chancel, L., T. Piketty, E. Saez and G. Zucman (2022), *World Inequality Report*, World Inequality Lab. Available online: https://wir2022.wid.world/ (accessed 4 October 2022).
Corak, M. (2016), 'Inequality from generation to generation: The United States in Comparison', *IZA Discussion Paper No. 9929*, Bonn: Institute of Labour Economics. Available online: https://www.iza.org/publications/dp/9929/inequality-from-generation-to-generation-the-united-states-in-comparison (accessed 4 October 2022).
Davies, M. (2013), *Oppositions and Ideology in Discourse*, London: Bloomsbury.
de Vries, R., S. Gosling and J. Potter (2011), 'Income inequality and personality: Are less equal U.S. states less agreeable?', *Social Science and Medicine*, 72: 1978–85.
Elgar, F. J., K. E. Pickett, W. Pickett, W. Craig, M. Molcho, K. Hurrelmann and M. Lenzi (2013), 'School bullying, homicide and income inequality: A cross-national pooled time series analysis', *International Journal of Public Health*, 58: 237–45.
Fairclough, N. (1989), *Language and Power*, Harlow: Longman.
Fairclough, N. (1992), *Discourse and Social Change*, Cambridge: Polity Press.
Fairclough, N. (1995), *Critical Discourse Analysis: The Critical Study of Language*, London: Longman.
Fairclough, N. (2002), 'Language in new capitalism', *Discourse & Society*, 13 (2): 163–6.
Fairclough, I. (2016), 'Evaluating policy as argument: The public debate over the first U.K. austerity budget', *Critical Discourse Studies*, 13 (1): 57–77.
Forsey, A. (2017), 'Hungry holidays: A report on hunger amongst children during school holidays', Report, All-Party Parliamentary Group on Hunger. Available online: https://www.basw.co.uk/resources/hungry-holidays-report-hunger-amongst-children-during-school-holidays (accessed 4 October 2022).
Fowler, R. (1991), *Language in the News*, Abington: Routledge.
Gabrielatos, C. (2007), 'Selecting query terms to build a specialised corpus from a restricted-access database', *ICAME Journal*, 31: 5–43.

Gabrielatos, C. (2018), 'Keyness analysis: nature, metrics and techniques', in C. Taylor and A. Marchi (eds), *Corpus Approaches to Discourse: A Critical Review*, 225–58. Routledge: Oxford.

Gale Historical Newspapers (2022). Available online: https://www.gale.com/intl/primary-sources/historical-newspapers (accessed 4 October 2022).

Gómez-Jiménez, E. (2018), '"An insufferable burden on businesses?" On changing attitudes to maternity leave and economic-related issues in the *Times* and *Daily Mail*', *Discourse, Context & Media*, 26: 100–7.

Gómez-Jiménez, E. and L. Bartley (2023), '"Rising number of homeless is the legacy of Tory failure": Discoursal changes and transitivity patterns in the representation of homelessness in *The Guardian* and *Daily Mail* from 2000 to 2018', *Applied Linguistics*, 44 (4): 771–90.

Gómez-Jiménez, E. and M. Toolan, eds (2020), *The Discursive Construction of Economic Inequality: CADS Approaches to the British Public Discourse*, London: Bloomsbury.

Graham, C. and B. K. O'Rourke (2019), 'Cooking a corporation tax controversy: Apple, Ireland and the E.U.', *Critical Discourse Studies*, 16 (3): 298–311.

Grisold, A. and H. Silke (2019), 'Denying, downplaying, debating: Defensive discourses of inequality in the debate on Piketty', *Critical Discourse Studies*, 16 (3): 264–81.

Halliday, M. A. K. and C. M. I. M. Matthiessen (2014), *Halliday's Introduction to Functional Grammar*, 4th ed, London: Routledge.

Hardie, A. (2014), 'Log-ratio – An informal introduction. ESRC Centre for Corpus Approaches to Social Science (CASS)', Available online: http://cass.lancs.ac.uk/log-ratio-an-informal-introduction/ (accessed 29 November 2022).

Hart, C. (2008), 'Critical discourse analysis and metaphor: Towards a theoretical framework', *Critical Discourse Studies*, 5 (2): 91–106.

Ho, D. (2022), *Notepad++* [Computer Software]. Available online: https://notepad-plus-plus.org/ (accessed 24 November 2022).

Iacoviello, M. (2008), 'Household debt and income inequality, 1963–2003', *Journal of Money, Credit and Banking*, 40 (5): 929–65.

Incelli, E. (2021), '"But what's so bad about inequality?" Ideological positioning and argumentation in the representation of economic inequality in the British press', *Lingue e Linguaggi*, 42: 77–100.

Jeffries, L. (2010), *Critical Stylistics*, London: Palgrave.

Jeffries, L. and B. Walker (2019), 'Austerity in the commons: A corpus critical analysis of austerity and its surrounding grammatical context in Hansard (1803–2015)', in K. Power, T. Ali and E. Lebdušková (eds), *Discourse Analysis and Austerity: Critical Studies from Economics and Linguistics*, 53–79. London: Routledge.

Kilgarriff, A., V. Baisa and J. Bušta (2014), '*The Sketch Engine*: ten years on', *Lexicography ASIALEX*, 1: 7–36.

Kondo, N., G. Sembajwe, I. Kawachi, R. M. Van Dam, S. V. Subramanian and Z. Yamagata (2009), 'Income inequality, mortality, and self-rated health: Meta-analysis of multilevel studies', *BMJ*, 339: b4471.

Lansley, S. and J. Mack (2013), 'A more unequal country?'. Available online: https://www.poverty.ac.uk/editorial/more-unequal-country (accessed 14 March 2022).

Lorenzo-Dus, N. and S. Marsh (2012), 'Bridging the gap: Interdisciplinary insights into the securitization of poverty', *Discourse & Society*, 23: 274–96.

Louw, B. (1993), 'Irony in the text or insincerity in the writer? The diagnostic potential of semantic prosodies', in M. Baker, G. Francis and E. Tognini-Bonelli (eds), *Text and Technology: In Honour of John Sinclair*, 157–75. Amsterdam: John Benjamins.

Machin, D. and J. E. Richardson (2008), 'Renewing an academic interest in structural inequalities', *Critical Discourse Studies*, 5 (4): 281–7.

Marquand, D. (2013), *Mammon's Kingdom: An Essay on Britain Now*, London: Allen Lane.

Mautner, G. (1995), '"Only connect." Critical discourse analysis and corpus linguistics', *UCREL Technical Papers 6*. Available online: http://ucrel.lancaster.ac.uk/papers/techpaper/vol6.pdf (accessed 14 March 2022).

Meinhof, U. H. and K. Richardson, eds (1994), *Text, Discourse and Context: Representations of Poverty in Britain*, London and New York: Longman.

Morrisson, C. and F. Murtin (2013), 'The Kuznets curve of human capital inequality: 1870–2010', *The Journal of Economic Inequality*, 11 (3): 283–301.

Mussolf, A. (2012), 'The study of metaphor as part of critical discourse analysis', *Critical Discourse Studies*, 9 (3): 301–10.

Newspapers.com (2022). Available online: https://www.newspapers.com/ (accessed 22 November 2022).

Nexis Uni (2022). Available online: https://advance.lexis.com/bisacademicresearchhome?crid=c55fecff-bc97-442c-92dd-2f61f3defd09&pdmfid=1516831&pdisurlapi=true (accessed 22 November 2022).

O'Donnell, M. (2016), *The UAM Corpus Tool 3.3*. Available online: http://www.corpustool.com (accessed 24 November 2022).

OECD (2022), *Income inequality data*. Available online: https://data.oecd.org/inequality/income-inequality.htm (accessed 4 October 2022).

Offer, A., R. Pechey and S. Ulijaszek (2012), *Insecurity, Inequality, and Obesity in Affluent Societies*, Oxford: Oxford University Press.

Office for National Statistics (2019), 'Household income inequality, UK: Financial year ending 2019', Newport: Office for National Statistics. Available online: https://www.ons.gov.uk/peoplepopulationandcommunity/personalandhouseholdfinances/incomeandwealth/bulletins/householdincomeinequalityfinancial/financialyearending2019 (accessed 4 October 2022).

PamCo (2022), 'UK newspapers brand reach September 2020 to June 2022'. Available online: https://pamco.co.uk/pamco-data/infographics/ (accessed 21 November 2022).

Partington, A. (2010), 'Modern diachronic corpus-assisted discourse studies (MD-CADS) on UK newspapers: An overview of the project', *Corpora*, 5 (2): 83–108.

Partington, A., A. Duguid and C. Taylor (2013), *Patterns and Meanings in Discourse: Theory and Practice in Corpus-Assisted Discourse Studies (CADS)*, Amsterdam: John Benjamins.

Paterson, L. L., L. Coffey-Glover and D. Peplow (2016), 'Negotiating stance within discourses of class: Reactions to benefits street', *Discourse & Society*, 27 (2): 195–214.

Paterson, L. L., D. Peplow and K. Grainger (2017), 'Does money talk equate to class talk? Audience responses to poverty porn in relation money and debt', in Mooney and E. Sifaki (eds), *The Language of Money and Debt*, 205–31. London: Palgrave.

Piketty, T. (2014), *Capital in the Twenty-First Century*, Cambridge: The Belknap Press of Harvard University Press.

Pryor, F. L. (2012), 'The impact of income inequality on values and attitudes', *The Journal of Socio-Economics*, 41: 615–22.

Rieder, M. and H. Theine (2019), '"Piketty is a genius, but … ": An analysis of journalistic delegitimation of Thomas Piketty's economic policy proposals', *Critical Discourse Studies*, 16 (3): 248–63.

Roberts, C. (2017), 'The language of "welfare dependency" and "benefit cheats": Internalising and reproducing the hegemonic and discursive rhetoric of "Benefit Scroungers"', in A. Mooney and E. Sifaki (eds), *The Language of Money and Debt*, 189–204. London: Palgrave.

Sandel, M. J. (2012), *What Money Can't Buy. The Moral Limits of Markets*, London: Penguin.

Silke, H., F. Quinn and M. Rieder (2019), 'Telling the truth about power? Journalism discourses and the facilitation of inequality', *Critical Discourse Studies*, 16 (3): 241–7.

Solt, F. (2010), 'Does economic inequality depress electoral participation? Testing the Schattschneider hypothesis', *Political Behaviour*, 32: 285–301.

Stiglitz, J. (2012), *The Price of Inequality*, London: Penguin.

Stubbs, M. (1996), *Text and Corpus Analysis*, Oxford: Blackwell.

Stubbs, M. and A. Gerbig (1993), 'Human and inhuman geography: On the computer-assisted analysis of long texts', in M. Hoey (ed.), *Data, Description, Discourse. Papers on the English Language in Honour of John Sinclair on His Sixtieth Birthday*, 64–85. London: HarperCollins.

The Equality Trust (2022), 'How is economic inequality defined?'. Available online: https://equalitytrust.org.uk/how-economic-inequality-defined (accessed 24 November 2022).

Toolan, M. (2016) 'Peter Black, Christopher Stevens, class and inequality in the *Daily Mail*', *Discourse & Society*, 27 (6): 642–60.

Toolan, M. (2018), *The Language of Inequality in the News: A Discourse Analytic Approach*, Cambridge: Cambridge University Press.

UK Press Online (2022). Available online: https://ukpressonline.co.uk/ukpressonline/ (accessed 22 November 2022).

van der Bom, I., L. L. Paterson, D. Peplow and K. Grainger (2018), '"It's not the fact they claim benefits but they're useless, lazy, drug taking lifestyles we despise": Analysing audience responses to benefits street using live tweets', *Discourse, Context & Media*, 21: 36–45.

van Dijk, T. A. (1995), 'Discourse semantics and ideology', *Discourse & Society*, 6 (2): 243–89.

van Dijk, T. A. (1998), *Ideology: A Multidisciplinary Approach*, London: Sage.

van Dijk, T. A. (1988), *News as Discourse*, New Jersey: Lawrence Erlbaum.

van Treeck, T. (2014), 'Did inequality cause the U.S. financial crisis?', *Journal of Economic Surveys*, 28 (3): 421–48.

7

Modelling the Landscape of Wilfred Owen's 'Futility'

Marcello Giovanelli

Introduction

In his study of embodiment in First World War poetry, Das (2007: 77) has argued that the defining characteristics of the genre are the stark movement away from epic forms and the refashioning of verse as 'missives from the trenches', both of which shift the emphasis of literary accounts of experience onto the interaction between body and landscape. This type of interaction appears to be well-suited to be studied through the lens of *ecostylistics*, which, as Zurru (2017: 195) suggests, may have one of two defining concerns:

> ecostylistics can focus either (a) on the link between the linguistic representation of physical environment and the style of a certain text, text-type, genre and/or author; (b) on the investigation and evaluation of (un)ecological linguistic patterns in texts, thus contributing to raising further awareness of global ecological concerns.

This chapter addresses the first of these two concerns by analysing Wilfred Owen's poem 'Futility' (Owen 1920), drawing attention to the way in which Owen represents the interaction between the body (here of a dead solider) and the landscape (the frozen battlefield of the Front in France). I argue that this interaction is the result of two specific kinds of modelling. First, as a poet, Owen models the consciousness of the observing speaker of his poem 'Futility', which is presented to the reader as a situated experience through the poem's grammar. Second, the reader models not only the situation but the distinctive minds presented within the poem, including that of Owen as the creative poet. Here readers draw on specific cognitive models of war, including biographical information, in order to present authoritative readings. In order to exemplify these points, I integrate my own cognitive stylistic analysis of 'Futility' with reader response data generated from online analyses of the poem. Overall then, this chapter maintains a dual focus both on the language used to describe the physical nature of landscapes and on the ways that readers articulate their experience of reading and understanding the poem (see Douthwaite et al. 2017).

The landscapes of the First World War

As Hynes (1990: 158) suggests, a turning point in First World War poetry arrived in 1916, when poems began to be published by soldier-poets, who presented the realities of war with 'the authority of direct experience', and whose verse was thus distinctive from the more idealistic and nationalistic war literature written by non-combatants.

One specific focus of this so-called trench poetry was directly concerned with representing the interaction between bodies and landscapes: bodies move into landscapes and are described within them; they display varying degrees of agency within them; they destroy or are destroyed by others within them; and landscapes themselves assume various literal or symbolic roles too that become body-like. *Landscape* in First World War literature is then a specific type of context, dynamically evolving and understood as a series of situations or 'features of the immediate physical surroundings that have significant meaning for some individuals (but not others) that define their psychological niche' (Barrett et al. 2010: 9).

It is no surprise, then, that landscapes become a defining feature of the genre, either explicitly or in more indirect ways. Here, for example, is part of Arthur Graeme West's poem 'God! How I Hate You, You Cheerful Young Men' (West 2007), written in response to early idealistic war verse.

> And *he'd* been to France,
> And seen the trenches, glimpsed the huddled dead
> In the periscope, hung in the rusting wire:
> Choked by their sickle fœtor, day and night
> Blown down his throat: stumbled through ruined hearths,
> Proved all that muddy brown monotony,
> Where blood's the only coloured thing. Perhaps
> Had seen a man killed, a sentry shot at night,
> Hunched as he fell, his feet on the firing-step,
> His neck against the back slope of the trench,
> And the rest doubled up between, his head

> Smashed like an egg-shell, and the warm grey brain
> Spattered all bloody on the parados:
> Had flashed a torch on his face, and known his friend,
> Shot, breathing hardly, in ten minutes – gone!

Initially, the human body is simply specified through the use of the third-person pronoun 'he', a common feature of war poetry (see Giovanelli 2014), and is spatially positioned relative to the French landscape, the trenches and 'the huddled dead' who, in turn, are 'hung in the rusting wire'. The poem dramatically portrays the sight of the 'sentry shot at night', foregrounds the landscape as a site of gruesome death, 'his head/Smashed like an egg-shell' before the soldier's realization that it is his friend, barely alive and, later, 'in ten minutes – gone!' The poem's lines iconically recreate the shifting nature of the landscape, from initial foreign soil 'France' to the graveyard of the sentry. The impact, of course, is keenly felt in the landscape itself, which is both the site of death but also is crucially affected by the action of those who fight within it: the building of the trenches, the ruined hearths, the blood which colours the land and, most starkly, the 'warm grey brain' of the dead solider which is splattered across the back of the trench so as to affect its very fabric.

Often interactions are mediated through touch, a physical force that binds human and landscape together in a symbiotic relationship. Das (2005: 8) argues that 'imaginative writing of the period repeatedly dwells on moments of tactile contact [...] gathered into the creative energies of a text', and claims that in Owen's poetry, bodies are thrust into or positioned carefully within their landscapes so as to both foreground the material itself and to emphasize the mediation between body and external world (2005: 153). We see this occurrence, for example, in possibly Owen's most famous poem 'Dulce et Decorum Est' (Owen 1920).

> Bent double, like old beggars under sacks,
> Knock-kneed, coughing like hags, we cursed through sludge,
> Till on the haunting flares we turned our backs,
> And towards our distant rest began to trudge.

Here the soldiers' bodies are foregrounded and afforded attentional prominence through their syntactic position in the first two lines, which outlines their physical characteristics through the adjectival phrases 'bent double' and 'knock-kneed', the noun phrase with a post-qualifying prepositional phrase 'old beggars under sacks', and the subordinate clause 'coughing like hags', all of which position the body as a primary focal point before the delayed main clause 'we cursed through the sludge' provides the background to the scene. By the time the landscape appears, it is thus presented as being an integral part of the soldiers themselves.

A further characteristic of First World War literature arises from the extreme physical conditions of weather, sickness and living conditions that those fighting had to endure. One of the most notable features of the landscape that became dominant in literary representations of the Front was the presence of mud and there are numerous examples of writers drawing on muddy conditions to highlight either the horror of the Front or as an extended metaphor for the war more generally (see Giovanelli 2022: 41–3 for discussion and examples). Most famously, Mary Borden's 'Song of the Mud', which originally appeared in the epilogue to her prose collection *The Forbidden Zone* (Borden 1929), emphasizes the ubiquitous nature of mud in a way that fuses the human to the landscape; in her poem, the mud is inescapable as it becomes 'the uniform of the poilu [infantryman]' and overcomes and brings to a halt the lives of the soldiers contained within it: 'wriggles', 'fills', 'mixes', 'crawls', 'spreads', 'sucks', 'soaks up'. Here, Borden's vocabulary choices focus on the actions of the mud as it becomes an agentive force as part of the landscape. In other instances, however, the emphasis is on the bodies of soldiers and their reconfiguration in the symbiotic relationship that they hold with the landscape. Das (2005: 44) articulates this relationship by outlining how the soldiers' body schema is realigned so that it now incorporates a sense of horizontality:

> the absolute lowering of the body on the ground allies seeing with the 'base senses' of touch and smell. The trench mud thus challenged the vertical organization of bodily Gestalt, and marked a regression to the clumsy horizontality of beasts. Rat, mole, earthworm and snail are recurrent similes that are used in trench narratives to describe the soldiers.

Examples of this realignment appear frequently in war accounts. Henri Barbusse's *Le Feu* (Barbusse 2003) contains multiple examples of soldiers crawling along the landscape of the Front, either across dead bodies, through the mud or simply in the dark. Barbusse's novel also emphasizes the transformation of the human body into a more animalistic gestalt. In the following example, the returning soldier is represented as devoid of a human form, instead taking on the physical attributes of a slug:

> 'The body of Mesnil Andre was not found, and his brother Joseph did some madescapades in search of it. He went out quite alone into No Man's Land, where thecrossed fire of machine-guns swept it three ways at once and constantly. In

the morning, dragging himself along like a slug, he showed over the bank a face blackwith mud and horribly wasted.

(Barbusse 2003: 209)

Another further common mediating weather element was the cold, and especially the snow and ice that came with being outside in the winter months. As Winter (1978: 95) outlines:

> Civilians contemplating trench war today would tend to think of it largely in terms of artillery and sniping action, raids and patrols. When the old soldier looks back over the years to his trench duty, however, he remembers clearly how seldom these actions interrupted the prolonged inactivity. To him, the real enemy was the weather and the side effects of living rough.

In the following extract (Barbusse 2003: 239), the soldier has died in the process of working on the embankment at the top of a trench.

> A body has slipped slightly to one side from upright, his chest and arms restingagainst the embankment. He was moving some earth when he died. His face, raised tothe sky, is covered with a leprous layer of rime. His eyelids are white and hismoustache covered with a hard spittle.

> Other bodies are sleeping, not as whitened as the rest; the layer of snow only staysintact on things: objects and the dead.

In this instance, the process of working within the landscape results in the solider becoming part of the landscape; death itself is the mediating force here.

The landscape of 'Futility'

Wilfred Owen's poem 'Futility' is a very good example of a poem that is concerned with the relationship and interaction between the human body and the landscape, here within the specific context of the cold weather. The poem is reprinted below.

Futility

Move him into the sun –
Gently its touch awoke him once,
At home, whispering of fields half-sown.
Always it woke him, even in France,
Until this morning and this snow.

> If anything might rouse him now
> The kind old sun will know.
> Think how it wakes the seeds –
> Woke once the clays of a cold star.
> Are limbs, so dear-achieved, are sides
> Full-nerved, still warm, too hard to stir?
> Was it for this the clay grew tall?
> – O what made fatuous sunbeams toil
> To break earth's sleep at all?
>
> (Owen 1920: 25)

Described by Corcoran (2007: 94) as 'Owen's greatest poem', 'Futility' was published (together with 'Hospital Barge') in *The Nation* in June 1918 and was one of five, and the last, of Owen's poems published in his lifetime. The poem was probably written during Owen's time at the Northern Command Depot at Ripon where he spent the spring of 1918 and is one of several poems he wrote that are set at dawn; Owen had initially planned for the poem to be published under the heading 'grief', one of several 'motives' that would provide a structure through his war poems (see Hibberd 2002: 316–8). 'Futility' provides a poignant account of the death of a soldier from exposure to the cold and the attempt by another to save his life and thus aligns with, but contrasts to, another poem 'Exposure' that captures the broader consequence of the cold weather on an entire platoon of men. Both poems are likely to have been influenced by Owen's own experience of extreme weather at the Front, as detailed, for example, in this letter to his mother in February 1917.

> My Platoon had no dug outs, but had to lie on the snow under the deadly wind. By day it was impossible to stand up or even crawl about […] The marvel is that we did not die of cold. As a matter of fact, one of my party actually froze to death before he could be got back, but I am not able to tell how many may have ended in hospital.
>
> (Owen 1985: 216)

'Futility' has been read as a stark rejection both literally to the touch of the fellow soldier (Das 2005: 160) and to different kinds of belief, from 'the sun's blessing' (Hibberd 2002: 122) to 'the pieties of orthodox Christianity' (Corcoran 2007: 94), to the 'metaphysical or philosophical framework in which they [soldiers and wider humanity] believe' (Macleod 1982: 240). It adopts a stance in which a speaker overlooks a scene, common in Owen's poetry, which Sokołowska-Paryż (2021: 390) describes as 'the role of a compassionate observer of the acts of war', and Castiglione (2020: 59) as concerned with the 'here and now of perception'.

My analysis that follows draws on Cognitive Grammar (Langacker 2008) and specifically the ways in which different kinds of processes are represented. Cognitive Grammar offers a way of analysing clause types through its notion of the 'action chain' (Langacker 2008: 355) that presents interactions between clausal participants as derived from a 'billiard ball' model where energy is distributed along a path from one entity to another. These clausal participants may fulfil several archetypal roles: an *agent* that carries out actions on other entities and is the source of energy; a *patient* which acts as the endpoint of the transfer or *energy sink* and, as part of the process, undergoes some internal change of state; and an *instrument* which is used by an agent and so plays an intermediary role in the action chain. Examples of these are provided below; the first represents what Langacker (2008: 357) terms the 'canonical event model', a prototypical clause in which an agent as energy source acts on a patient as energy sink to bring around some kind of change. This prototype is also diagrammatically modelled in Figure 7.1.

1. I (agent) kicked the ball (patient)
2. I (agent) kicked the ball (patient) with my left foot (instrument)
3. I (agent) kicked the ball and the other player (double patient)
4. I (agent) kicked myself (agent also patient)

Other archetypal roles occur in single-participant processes that do not model transfers of energy, what are known as thematic processes (Langacker 2008: 370). These include an *experiencer*, a sentient human entity who is the primary participant in some kind of mental or emotional process involving a

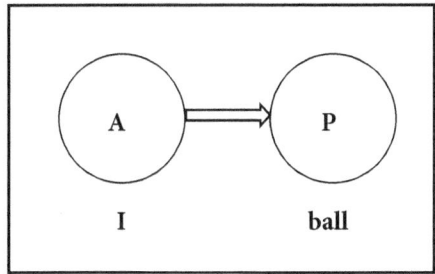

Figure 7.1 The canonical event model.

stimulus (e.g. He loved the cake), a *mover*, either an animate or inanimate object which shifts location through its role as a participant in a process (e.g. The fence blew over), and a *zero participant*, characterized by its stasis within the clause (e.g. The wall was high).

The first line of the poem begins with an imperative 'Move him'. Here the agent of the action chain is implied but not explicitly stated; the attention thus falls on the process itself and the patient, understood as being the dead solider. The speaker is unnamed and, in Langacker's stage model (Langacker 2008: 356–7), which outlines further archetypes for the viewing arrangement of a scene, remains 'offstage' since the focus of attention is on the scene being described rather than the viewer or act of viewing itself. As the poem moves into line 2, this clause pattern is reconfigured so that the location/setting of the sun that marks the endpoint of the initial action chain (conveyed through the prepositional phrase 'into the sun') assumes the status of clausal subject and the role of agent. Now the verb 'awoke', used transitively, profiles an action chain in which the (implied) dead solider still remains as patient, but on this occasion, it is the sun's energy which fully wakes him. The temporal parameters of the waking are specified using the adverb 'once', which invokes a particular kind of 'scanning scenario' (Langacker 2008: 532) in which a virtual instance (here the waking of the soldier) is understood as standing in for multiple actual instances.

The reversal in agent-patient roles depends, of course, on our understanding of the temporal shift enacted through the change to the past tense so that we understand the situation being described in contrast to the current space in which the speaker – and the sun – are unable to wake the fellow soldier. The use of the frequency adverb 'Always', which places further attentional prominence on the activity works, much like 'once' two lines earlier, to 'prompt the simulation of a scanning experience' (Langacker 2008: 533) since we are not provided in the poem with every instance that the sun woke the soldier but are reminded, through a virtual instance, of the sun's power to sustain life. The optimism afforded by this particular construal, however, diminishes in the remainder of the stanza as the poem's grammar positions the reader to adopt another shift through the use of the proximal determiners 'this morning ... this snow' and the temporal adverb 'now', which ground the poem's action both temporally and spatially in the more specific frame of the immediate battlefield in France.

The degree of specificity inherent in the determiners is in contrast with the nondefinite pronoun 'Anything', which assumes the role of agent, with the soldier as patient in the penultimate line of the first stanza. 'Anything' also profiles a virtual entity, although unlike 'once' and 'Always', its use seems to suggest less certainty in the projected instance that is being profiled, evident, I think, in that it follows the line 'Until this morning and this snow'. The effect, it seems to me, is to downplay the potential transfer of energy from source to sink and thus provides a turning point from the more optimistic sense of the previous lines. The real-time viewing of the poem is, therefore, a recognition of the inevitable death of the solider and the inability of the sun (or anyone/thing else) to replicate the various instances of awakening that have happened before.

Further textual evidence for this reading appears in the form of the modal pairing 'might' and 'will', the playful adjectival modification of 'kind old sun', and the final thematic process in which the sun is the experiencer, all of which draw attention away from any energy transfer and thus downplay the sun's agency. Together the patterning of clauses across the first stanza profiles a set of actions in which an energy transfer is first demanded, then identified in one single instance, then generalized across many unspecified instances, and then (weakly) projected as a future instance. The poem's grammar positions the reader to move from expectancy to hope to a lack of certainty. The fact that the final clause profiles a thematic process removes human agency completely when compared to the initial addressee of the imperative 'move'. The first stanza thus begins by profiling an energy transfer and ends by emphasizing stasis. Overall, then, the grammar works to define the soldier's relationship with the landscape as follows:

1. Soldier is moved within landscape
2. The landscape (sun) is asked to act on soldier
3. The solider is imagined as positioned within an alternative landscape – at home
4. The landscape of France is unable to sustain the soldier, unlike the landscape of England (and earlier iterations of France)
5. The stanza ends with a sympathetic plea, largely an acceptance of the failure of landscape to maintain life and sustain the soldier.

The second stanza also begins with an imperative but on this occasion, the verb 'think' positions the unknown addressee as the experiencer in a cognitive process. Here the addressee is likely to be understood as a more general one than the specific entity who acts as the implied agent of the action chain in the first line of the first stanza. The secondary entity in this thematic process contains two embedded action chains with the sun as agent/energy source and the 'seeds' and then 'the clays of a cold star' as patients/energy sinks. The reader is positioned to reflect on the more philosophical aspect simply because these action chains now point beyond the immediate scope of the situation depicted in the first stanza to give a sense of broader significance. First, although the initial use of the present tense 'wakes' imposes an immediate scope (the 'onstage' part of the action) that coincides with the time of the speaking voice, the duration of the action far outlasts the telling of it and so invokes a more schematic representation of the situation. Second, the shift, marked by 'woke', positions the temporal frame to well before the situation time of the poem and so the immediate scope is profiled as independent of the poem's events themselves (interestingly here the past tense and use of the indefinite adverb 'once' also mirror a pattern present in the first stanza).

The poem's grammar is now less direct in how it specifies agency through the rhetorical questions that complete the stanza. The first, beginning 'Are limbs', is marked as a non-finite infinitival clause, again with an unspecified agent for the verb 'to stir'. The second is more explicit in its removal of agency, consisting of an embedded process in which the 'clay' (earth) is profiled as a mover in a process with no hint of an action chain. And the third, which ends the poem, starts with another unknown agent 'what', before the patient 'fatuous sunbeams' assumes the role of agent of the verb 'to break' in the final clause, with 'earth's sleep' as patient. In this instance, the cohesion is understood in Cognitive Grammar terms as a series of reference point relationships (Langacker 2008: 504) in which one entity (an initial reference point) provides mental access to another (a target which then becomes a reference point providing access to further targets) within its dominion (the associated knowledge or co-text). Here then, the cohesive chain extends across the poem from stanza one to stanza two: kind old sun≫it≫seeds≫cold star≫limbs≫clay≫sunbeams (see also Stockwell 2009: 179–82 for a discussion of this phenomenon in Dickens' *Bleak House*).

Overall, the organization of the second stanza is such that there are less explicit action chains with a greater number of thematic processes present. But another clear pattern is that human agency is downplayed as the solider (and I would argue the poem's speaker) fades into the attentional background and a new figure of the sun/landscape itself emerges. That this focus centres on a broader philosophical question is, of course, important. It seems to me that the poem progressively pushes readers outwards from the immediacy of the opening image; its organization is such that each process acts as part of a broader attentional frame that increasingly shifts the figure-ground configuration so that the landscape itself becomes the vehicle for the grander philosophical question that Owen's speaker appears to want us to consider. In this case, the turning point of the poem is the shift from first to second stanza (at odds with the traditional *volta* of a sonnet that occurs after line 8). This turning point may also be read as a type of radical reconstrual (a type of re-presentation) and specifically a *refiguring* of attention (Giovanelli 2022: 154) that shifts the focus from the individual to the metaphysical and existential.

I want to end my analysis by returning to the speaker of the poem and to the consciousness that Owen models. A striking feature of the poem is the absence of the first and second pronouns 'I' and 'you'; even when imperative forms and questions are used, neither the speaker or hearer are placed 'onstage' and given direct attentional prominence. Yet, for me at least, this is a poem in which the subjectivity – and the pain – of the speaking voice do resonate and it seems that Owen is particularly effective here at creating a speaking voice situated within a specific context. It also seems to me that one of the primary characteristics of the poem is its ability to create, through its various attentional frames and the processes embedded within them, a very clear and significant message about the war itself in that the event is conceived as powerful enough to first give rise to the soldier's pleas and hopeful optimism, and then to the broader philosophical querying and ultimate sense of despair. In other words, the whole poem becomes framed as a powerful *situation*, defined by Barrett et al. (2010: 9) as 'not a description of the physical properties of the environment, but … characterized as containing just those aspects that are relevant to the thoughts, feelings and behaviors of that particular person at that particular point in time'.

This seems a neat way of capturing Owen's success in 'Futility'; the poem models an ongoing consciousness that uses the landscape of the Front to both precisely and poetically frame the personal significance of a soldier's death, and to examine more broadly the human cost of war.

Interpreting 'Futility'

The final section of this chapter examines some actual responses beyond my own to Owen's poem. These are provided both to test some of the claims made by the application of Cognitive Grammar and to sit alongside my own interpretations as ways in which readers draw out the significance of the landscape in 'Futility' and model the consciousness of Owen's speaker. The data I draw on consist of a small corpus of online summaries of the poem, produced to be used primarily to educate readers, either students or just those with a general interest, about Owen's poem.

Whereas most online reader response data come from book review sites, personal blogs or discussion boards, my choice to use these more formal analyses is based on the fact that these tend to present detailed interpretations, do so in an authoritative manner and have the potential to be influential in terms of presenting the meaning of a text thus presenting more stable interpretations beyond the simply idiosyncratic. Thus, although they largely present less personal accounts of literary readings, they do offer an interesting data set since they tend to frame readings (and readers) in particular ways. The websites used are the first eight sites that appeared in a search on Google using the term 'Wilfred Owen Futility analysis' (excluding those which required paid subscription for full access), which matched the criterion of being produced primarily for educational purposes. The corpus was managed and inductively coded in NVivo to draw out emergent themes, which are discussed in the remainder of this section.[1]

Landscapes do feature prominently in the online analyses. Often, some sense of the landscape is presented in the initial summary of the poem: for example, '"Futility" follows the aftermath of a battlefield' (W1). In most cases, however, the landscape of the poem is presented in a more negative sense. W7 comments that 'All the creative and stimulative forces of Nature [...] stand paralysed and pathetically helpless'. Similar points are made by W8, which draws on the contrast between the 'Futility's two stanzas where initially 'Owen, returning to his Romantic roots, admires the power of nature, the power of the sun' outlining 'how a deceased soldier is moved to the sun with the hope that the gentle rays of the sun will revive his consciousness'.

Further comments, however, outline the relationship of the landscape to the acceptance of death. Although not explicitly stated in the poem, W8 reads the landscape of the poem as being 'devastated' and argues that 'nature is a passive, everlasting witness of the destruction wrought by mankind', while W1 argues that, in the poem, 'Owen draws a connection between life, like the soil, and the man, now devoid of it'. These comments align with my analysis of the poem which presents the landscape as increasingly attentionally significant whilst at the same time questioning its agency.

In the same manner, many of the websites draw readers' attention explicitly to the sun as an integral part of the landscape that Owen describes in 'Futility'. W3, for example, suggests that the sun is the 'metaphorical framework on which to hang his [Owen's] thoughts'. In many cases, the websites specifically comment on the disconnect between what the sun is conventionally able to achieve and what is denied in the context of the dead soldier in the poem. The following comment mirrors, I would suggest, the grammatical patterning of the poem in which action chains presenting energy transfer are either replaced by more thematic processes or else framed as unable to be realized.

> The sun wakes us (lines 2 & 4), stimulates us to activity (3), holds the key of knowledge (7), gives life to the soil (8), gave life from the beginning, yet (13) in the end the 'fatuous' sunbeams are powerless.
>
> (W3)

Consequently, the lack of agency is often identified as a central concern of the poem:

> The sun shines today as usual with its rays, touching the dead-body of the youngsoldier, but it is powerless to make him alive and active again.
>
> (W7)

> The sun is reliable and powerful, waking the soldier 'even in France', a foreign country far from the safety of his home. Yet this reliability makes the sun's inability to wake him now even more striking and frustrating. If the sun could always wake him before, why can't it do so now?
>
> (W4)

Indeed, the theme of helplessness more generally is prominent in the websites' interpretations of the poem, often framed in a way that suggests a sense of movement through the poem, from hope to despair. For example, one comment in W1's is 'Thus the first stanza ends on that lingering trace of hope – hope that is now dashed'; while W3 claims that 'Line 12's "Was it for this the clay grew tall?" has life, in man, reaching its peak merely to come to nothing, and the poem ends, fittingly, in ambiguity', and W8 'The second stanza depicts a change in tone of the speaker who takes on a questioning attitude regarding life'. Equally, the specific characteristics of the second stanza, described in my analysis as the move towards a more objective, but no less painful, existential questioning, are highlighted across a number of instances in my data set, such as:

> If the speaker of the poem really does begin the poem believing that the sun will be able to rouse the dead man, it may be that the questions which appear in the second stanza are genuine, asked through disbelief and a growing disillusionment.
>
> (W6)

Futility' ends on the silence that follows, leaving the questions unanswered and extinguishing all the sense of building hope that Owen has gently grafted throughout the poem. There is no answer. There is nothing, Owen seems to be saying, but blood and senseless death.

(W1)

My data set also reveals some interesting ways that biographical details are used to provide an interpretative frame for the poem. These range from outlining straightforward biographical facts to the following more nuanced examples of mind-modelling (Stockwell 2009), where an authorial presence is imposed on the poem so as to ascribe some meaningful motivation behind the various linguistic choices made (in all instances, emphases are my own).

This was a reality **known all too well** to the poet – young men were being killed before their lives had barely begun.

(W2)

Good or bad, the immense strain put on Owen by pushing him to lead the charge contributed to his poetry, as well as to **the growing sense of misanthropy that he suffered** as soon as he had returned to war.

(W1)

Owen subverts the trope by applying it to a soldier, and while scholars who point out Owen's sense of latent homoeroticism in his poems are not wrong, one should also consider **the closeness that Owen felt towards his fellow brothers-in-arms**.

(W1)

Owen **seems to be questioning** the need for human life – what is the point in leading such a paradoxical existence where life invariably leads to death?

(W8)

Owen's reference to light can be understood **in terms of his religious inclination** also. By light, he projects God, the life giver.

(W8)

As Stockwell (2016: 149) notes, readings which purport to have some access to authorial lives are frequently observed in both mainstream literary criticism and in the responses of lay or 'civilian' readers. This intentional stance can be explained as a set of inference-making procedures that a reader undertakes to make connections between the language of the text, its interpretative effects and the real-life author, a version of which is modelled in the creative act of reading and the search for a satisfactory and coherent meaning. Available information, therefore, about an author becomes significant and drawn on alongside other types of knowledge to produce a satisfactory reading. For First World War poets, this type of mind modelling might appear to be inescapable (see Stockwell 2016 on Isaac Rosenberg and Giovanelli 2022 on Siegfried Sassoon for similar discussion). For Owen, I would argue that his status as a canonical war poet, complete with the various cultural associations – at least in the UK – is so powerful and so pervasive that it is impossible to ignore matters of biography. This cognitive model for Owen as an anti-war poet therefore provides an interpretative frame for understanding the overarching way that the physical forms of the human landscape interact in 'Futility'. In broader terms, it seems that war poetry, and specifically Owen's own verse, is established as very specialist genres, comprising, in Cognitive Grammar terms, a set of recognizable schemas (Langacker 2008: 478) that are used by readers to articulate responses to the poem.

Conclusion

In this chapter, I have outlined how an understanding of the ways in which Owen presents the interaction between body and landscape in 'Futility' can be enriched by drawing on Cognitive Grammar to outline grammatical patterns and shifts across the poem's two stanzas and the ways in which these may position readers both to respond to the poem's central concerns with war, and to model and respond to the experiencing consciousness of the poem, conceived as existing within a profoundly powerful situation. This chapter thus provides an example of ecostylistic analysis, in Zurru's (2017) terms, by connecting the linguistic representation of the physical environment and the style of 'Futility' but broadens out the scope of such an analysis to consider the wider cognitive dimension in its modelling of a particular mind. More generally then, this chapter exemplifies the use of linguistic theory to explore how a 'text comes to convey a certain stance' (Zurru 2017: 195), highlighting how this stance is not just textually projected but also realized in responses to the poem which are presented as authoritative and stable. These responses align with my own examination of the poem in identifying and arguing for the importance of the landscape in 'Futility', and draw on specific extra-textual resources to articulate cognitive models of war over one hundred years after the end of the conflict.

Note

1 In terms of ethical consideration, the sites that I am drawing on are all publicly available and, unlike personal blogs, have a clearly stated educational purpose. They are clearly designed to be used as literary-critical material and, consequently, I treat them as similar to sources that might be found in academic books, journals, and on a variety of online academic/educational spaces. The eight websites that I use in this chapter are

W1: https://poemanalysis.com/wilfred-owen/futility/ (accessed 30 December 2022)
W2: https://www.gradesaver.com/wilfred-owen-poems/study-guide/summary-futility (accessed 30 December 2022)
W3: http://www.wilfredowen.org.uk/poetry/futility (accessed 30 December 2022)
W4: https://poetryshark.wordpress.com/2016/02/29/futility-by-wilfred-owen-poem-analysis-gcse/ (accessed 30 December 2022)
W5: https://literarydevices.net/futility/ (accessed 30 December 2022)
W6: https://interestingliterature.com/2015/10/a-short-analysis-of-wilfred-owens-futility/ (accessed 30 December 2022)
W7: https://literaryocean.com/explain-the-title-of-the-poem-futility-by-wilfred-owen/ (accessed 30 December 2022)
W8: https://smartenglishnotes.com/2021/02/16/summary-and-analysis-of-futility-owens-poetic-epitaph/ (accessed 30 December 2022).

References

Barbusse, H. ([1916] 2003), *Under Fire*, trans. By R. Buss, London: Penguin.
Barrett, L., B. Mesquita and E.R. Smith (2010), 'The context principle', in *The Mind in Context*, 1–24. New York/London: The Guildford Press.
Borden, M. (1929), *The Forbidden Zone*, London: Heinemann.
Castiglione, D. (2020), 'The stylistic construction of verbal imagery in poetry: Shooting distance and resolution in Wilfred Owen, Marianne Moore and Philip Larkin', in J. Piątkowska-Brzezińska and G. Zeldowicz (eds), *Znaki czy nie znaki? Tom 3. Struktura i semantyka utworów lirycznych* (*Signs or Not Signs? Volume 3. The Structure and Semantics of Lyric Works*), 43–79. Warszawa: Wydawnictwo Uniwersytetu Warszawskiego.

Corcoran, N. (2007), 'Wilfred Owen and the poetry of war', in N. Corcoran (ed.) *The Cambridge Companion to Twentieth-Century English Poetry*, 87–104. Cambridge: Cambridge University Press.

Das, S. (2005), *Touch and Intimacy in First World War Literature*, Cambridge: Cambridge University Press.

Das, S. (2007), 'War poetry and the realm of the senses: Owen and Rosenberg', in Tim Kendall (ed.), *The Oxford Handbook of British and Irish War Poetry*, 73–99. Oxford: Oxford University Press.

Douthwaite, J., D. Virdis and E. Zurru (2017), 'Introduction', in *The Stylistics of Landscapes, the Landscapes of Stylistics*, 1–20. Amsterdam: John Benjamins.

Giovanelli, M. (2014), 'Conceptual proximity and the experience of war in Siegfried Sassoon's "A working party"', in C. Harrison, L. Nuttall, P. Stockwell and W. Yuan (eds), *Cognitive Grammar in Literature*, 145–60. Amsterdam: John Benjamins.

Giovanelli, M. (2022), *The Language of Siegfried Sassoon*, Cham: Palgrave Macmillan.

Hibberd, D. (2002), *Wilfred Owen: A New Biography*, London: Weidenfeld and Nicolson.

Hynes, S. (1990), *A War Imagined: The First World War and English Culture*, London: The Bodley Head.

Langacker, R.W. (2008), *Cognitive Grammar: A Basic Introduction*, Oxford: Oxford University Press.

Macleod, N. (1982), 'The stylistic analysis of poetic texts: Owen's "Futility" and Davie's "The Garden Party"', in J. A. Anderson (ed.), *Language Form and Linguistic Variation*, 239–76. Amsterdam: John Benjamins.

Owen, W. (1920), *Poems by Wilfred Owen*, edited by S. Sassoon, London: Chatto and Windus.

Owen, W. (1985), *Selected Letters*, edited by J. Bell, Oxford: Oxford University Press.

Sokołowska-Paryż, M. (2021), 'Wilfred Owen, war poetry', in Ralph Schneider and Jane Potter (eds), *Handbook of British Literature and Culture of the First World War*, 381–96. Berlin and Boston: Walter De Gruyter.

Stockwell, P. (2009), *Texture: A Cognitive Aesthetics of Reading*, Edinburgh: Edinburgh University Press.

Stockwell, P. (2016), 'The texture of authorial intention', in J. Gavins and E. Lahey (eds), *World Building: Discourse in the Mind*, 147–64. London: Bloomsbury.

West, A.G. [1919] (2007), *The Diary of a Dead Officer: Being the Posthumous Papers of Arthur Graeme West*, edited by N. Jones, London: Greenhill Books.

Winter, D. (1978), *Death's Men: Soldiers of the Great War*, London: Allen Lane.

Zurru, E. (2017), 'The agency of *The hungry tide*: An ecostylistic analysis', in J. Douthwaite, D. Virdis, and E. Zurru (eds), *The Language of Landscapes*, 191–231. Amsterdam: John Benjamins.

8

Intralingual Eco-Translation Insights into *Macbeth* in African American Urban Slang

Michał Garcarz

Introduction

Translation seems to be a subtle form of an intercultural communicational hoax. One person (sometimes a group of those collaborating as a team) retells an original story in another language with his/her/their own words, paying utmost attention to offering the target language audience a translation – an optimal variant of the original which would steal their hearts and minds far greater than the genuine original could have ever done so. An interlingual translation act theoretically bridges cultural and gnoseological gaps between/among individuals thriving in systematized organisms – societies where there are no natural intercultural communication mechanisms ready for use to efficiently send/receive a text message across the mentioned cultural gaps. Furthermore, it is clear to everyone who has at least once suffered the need to perform a communication act with another person speaking exclusively his/her own native language that no other means of efficient intercultural communication is far more desired than the act of translation. Such a need grows with intensity if the text author desires the text message addressee to comprehend its sense, especially when the subject of the message itself takes the form of an artistic artefact.

    This chapter is another voice in the ongoing debate on the aesthetic and communicational values of the act of retelling in the same language a story once told, i.e., on the functional nature of an intralingual adaptation as an act of the interpretational challenge of the original semantic input. My aim here is to provide the reader with a contrastive analysis of stylistic operations that have been introduced to, if not 'imposed on', the Shakespearean *Macbeth* to 'retell' it in the present-day African American urban slang. Tonia Lee, who has adapted the original sense into a discursively alien tongue, has answered the Jakobsonian (1959: 223) call to perform the act of translation in another communicational dimension, which is an intracultural platform of thought exchange. From an eco-linguistic viewpoint, Tonia Lee's adaptation of *Macbeth* is indeed forceful when it comes to unmasking the discursive potential of the original being exposed to low-register communication practices typical of contemporary African American 'street speech' (Garcarz 2013: 184–93). The local-scale environmental eco-linguistic study of Tonia Lee's intralingual translation artefact has been planned to give the readers of this chapter answers to two major research questions:

- What translation tools – i.e., translation techniques – has Tonia Lee applied to her professional translation endeavour to achieve the African American urban street speech flavour in her adaptation of *Macbeth*?
- What are the stylometric differences between the two renderings of *Macbeth*, basing the research on a qualitative corpus analysis?

Answering these above-formulated research questions will give grounds to confirm or deny an overall assumption that an intralingual adaptation of a literary text is not an act of communication impossibility and that the eco-translation strategy of domestication is rhetorically effective.

Shakespearean *Macbeth* in a 'foreign' modern tongue

The English language world of cultural artefacts has always been dominating across the globe since the turn of the seventeenth century. When we look at the past three hundred years of the Anglo-Saxon significance in the world in the sphere of culture, literature and art, it should not be strange for anybody that peoples of other languages eagerly desired to learn about the reasons that made the English-speaking cultures attractive and catchy. Without translators and the services they have rendered to popularize the English language cultural assets, the Shakespearian heritage, for instance, would never have been recognized today in almost all pockets of the world. The act of translation is somehow naturally associated with a series of complicated interlingual operations on text with the view to mirror the original essence in the minds of the target language version readers. Intralingual translation is thus taken as a form of textual refreshment only, as a less serious, less time- and effort-consuming type of trivial play on words dedicated to the speakers of the same language set in later generations whose socio-cultural circumstances of everyday existence have (sometimes radically) changed.

My understanding of intralingual translation as a communication activity is far more different from the one presented above, especially when it comes to recreating a literary piece that is linguistically complex and artistically demanding, such as the Shakespearian *Macbeth*. First of all, the English language has changed a lot since the times of Elizabethan England; the lexicon along with the grammar rules is today used differently from what it was five hundred years ago. Today's English language speakers may have substantial difficulties comprehending the meaning of some dialogue/monologue turns when dealing with the original and really appreciating the artistic value of the whole text. A modernized version of such literary artefact brings then the today's audience closer to the original essence when its surface structure, i.e., the text body itself, has been adapted to their everyday language communication habits. I reckon Tonia Lee has successfully achieved that communication goal. Tonia Lee, Teacher's College, Columbia University graduate of 1994, educator at NYC Department of Education, teacher and social activist, has mirrored the Shakespearean spirit in a modern tongue of twentieth-century America to reach out to the same audience, only set in a different time. Moreover, this is the reason why Lee has decided to use contemporary urban slang of American multicultural neighbourhoods of New York City to enhance the social impact factor of her intralingual adaptation radically and to intensify possible effects of her educational mission involving young people. Yes, the educational purposes are what Tonia Lee has underlined as the primary objective of her translation activity: to reinforce young people's reading and learning activities. '*Macbeth* in urban slang condenses and modernizes Shakespeare's original *Macbeth*. It promises to be a sensational read for any teen, preteen, or person young at heart that either identifies with American urban pop culture, or is curious about it', points out Lee (2008: iv).

Having the above-stated references, we can now clearly see what is the major purpose of the appearance of intralingual translations in the public domain: to teach the next generations about our cultural ancestry, to popularize an art of text interpretation and to induce in the minds of the youth the habit of reading literature. These assumptions seem to remain compatible with Lee's (2008: iv) thoughts regarding the educational usefulness of her own translation:

> This book would be ideal for: An urban classroom, A classroom library, Reader's theater unit, A literary circle, An after school program, A book club for youth, A drama club or class, A gift, A parent to use to enrich his or her child's education,

A summer reading program, A person who is interested in reading a unique version of *Macbeth*.

In this case, it is fair to claim that Lee's intralingual operations on *Macbeth* fall into that type of translation strategy that Schleiermacher ([1813] 1973) termed 'domestication' (German *Einbürgerung*). A domesticated text is adapted to the target language audience's communication principles and everyday language contact routines. Promoted in the English-speaking world by Venuti (1995: 7–15), Schleiermacher's ([1813] 1973) binary classification of translation strategies, useful especially in literary translation practices, i.e., 'domestication' (German *Einbürgerung*) and 'foreignization' (German *Verfremdung*), gives the translator – in this case the literary text adaptor – a solid ideological foundation to systematically manufacture the original surface layer only according to his/her way of portraying the original essence in another tongue. Thus, if we assume that literal or word-for-word translation operations effect in the general simplification of the original (Guo and Wan 2022), only an adaptation-oriented translation strategy, that is, 'domestication' (Schleiermacher [1813] 1973), allows the translator to 'open up' the original essence in the act of translation. Tonia Lee's *Macbeth* is low-register lexis heavily laden. Such word stock carries the label of an idiolect, especially if the speakers who habitually use it constitute a speech community associated with a given variety of language (see Bauer 2002).

Bo (2014: 709) is of the opinion that 'eco-translation is a form of adaptation, to translate the original text into a new one'. Intralingual adaptation procedures result in changing the social parameters of text reception performed within the same language, which is credible evidence of eco-translation. Elizabethan England – which was a sociological cradle for *Macbeth* – has been swept for the streets of the African American youth living in contemporary NYC. Thus, a question arises as to what is the sociolinguistic specificity of the African American street speech that helped Tonia Lee successfully adapt *Macbeth* into another socio-cultural dimension of the English language world? It is the speakerhood of African American urban slang.

African American urban slang speakerhood

The street speech has been considered an effective means of projecting ethnic issues via platforms associated with African American Diaspora, such as Raping or MCing, and, in this way, the street speech adheres to the actual pronunciation and grammar of African American Vernacular English. Consequently, general

English language users may regard 'Hip Hop language' as confrontational (Smitherman 1998: 214). 'Street Speech was considered to be either different or deficient when compared to standard English'; that is Baugh's opinion (1983: 11), which seems justified when we consider the habitual orality of street speech. This is another logical and practical argument for choosing this sector of the informal register lexicon of the American English language to mirror the Shakespearean content in one of its present-day adaptations. The original *Macbeth* was too manufactured for a stage performance, which means the story itself used to be delivered orally. Therefore, orality is a primary property of the street speech, as is 'slangity'.

'Slang is more characteristic of urban Black talk because of the heightened importance of peer-grouping in the *hip* city environment' (Abrahams 1967: 20), and this is why the Hip Hop street speech holds firm relations with African American ghetto roots. For Green (2010: xiv), the slang typical of the 'streets' 'is the gutter language'. More on the sociolinguistic spread of urban African American Vernacular English is revealed by the research carried out by Cukor-Avila and Bailey (1996), which seems to confirm Smitherman's (1977: 50) opinion on AAVE that lexical and grammatical solutions as well as phonological patterns habitually practised by African Americans in their everyday conversations fall into the layer of American English vernacular she terms as 'Black Semantics' (Smitherman 1977: 50). Moreover, the 'Black Semantics' genotype is most often isolated in the course of a lexicographical analysis, when a sample of selected lexical units has been studied to categorize linguistically the spelling alterations that the AAVE lexicon in question carries.

There are five most prominent phonological characteristics of AAVE impacting the promotion of orthographic subvariants of numerous lexical units well recognized by the American English users that are typical of the street speech (and that should be associated with Black Semantics, I reckon): (1) 'monophthongization' (Rickford 1998), 'deletion of post- and intervocalic /r/' (Labov 1972), 'fricative stopping of [ð] and [θ] to [d] in word-initial position' (Rickford 1998), 'g-dropping' (Green 2002) and 'final consonant cluster reduction' (Labov 1972). Moreover, Lee's adaptation of *Macbeth* is rich in the stereotypical instances of Smithermanian Black Semantics, e.g., 'gangsta' or 'sista' ('But Macbeth! Yo, that dude is gansta!' [Act 1, Scene 2] or 'First Witch: Yo, where have you been, Sista?' [Act 1, Scene 3]). The stylistic variation of Lee's adaptation of *Macbeth* ranges from formal ('Macduff: No, I think I know what you are going to tell me'), through informal/semi-slangy ('Ross: I'm afraid so. Sorry <u>to break it to</u> you like this, man' [underlining by M.G.]), to fully slangy

('Malcolm: <u>Suck it up</u>! Take it like a man. That's what I had to do when it happened to my Pops.' [underlining by M.G.]). These examples extracted from one segment of the drama (Act 4, Scene 3) are situated literally on a single page of this book edition (i.e. page 45)! Knowing this, to evaluate the efficiency of the exploitation of the most effective eco-translation strategy – which is 'domestication' (see Schleiermacher [(1813) 1973]) – in Lee's adaptation endeavour, I, first, should give a more focus on what is communicational functions of slang.

Slang in a sociolinguistic view

Slang conceptually lies in the lowest language register, and it is always stylistically unstable; it is a vehicle of language modification and melioration processes on its colloquial spectrum of usage, mainly via speech (less often via writing). Moreover, slang generates social interactions and is socially dependent. Speech communities are associated with particular language communication rituals performed with the use of slang lexicon. All in all, slanging speech communities are of not highly insular nature, taking it into account from the sociological viewpoint; the fluctuation in the number of members of such slanging communities is high and utterly everyone applying slanguage for everyday communication is, in fact, the group's member in his/her full socio-communication rights. Therefore, slanguage, the less hermetic variant of slang, is sociolinguistically located on the borderline of colloquial speech and the slang itself. 'I translate slanguage into Street Speech – a public/personal sphere of oral communication – typical of informal or very informal situations when formalized and systematized language norms have no application, and when the borderline between colloquialisms and slang items is blurred or has faded away,' concludes Garcarz (2013: 112). Such is the urban slang of Lee's *Macbeth*. It stylistically ranges from the casual register, through colloquial to slangy, as I have presented in the previous part. Urban slang is truly stylistically unstable and covers all prototypical components of general slang, such as swear words, vulgarisms, tabooed language items, sexualisms, euphemisms, argot items and cant items (Garcarz 2013: 130–52). However, its word stock collection awaiting recipients of Lee's adaptation of *Macbeth* does not fall into all systematized components of the slang lexicon in its general sense. The intralingual eco-translation version of *Macbeth* lacks swear words, vulgarisms, sexualisms, argot items and cant items, as urban slang lexical categories. The permanently changing stylistic identification of urban slang in its figurative application to everyday communication is still an optimal instrument of thought exchange, chiefly thanks to one rhetorical tool: *compléments cognitifs*.

Compléments cognitifs is a cultural-language symbol reflecting speakers' knowledge of their cultural surroundings and their experience in interpersonal communication, which, when applied in intralingual communication, shrinks the time span of information exchange in a communication act. Individuals set in a communication act use common cultural knowledge triggered by learnt social connotations. These give all individuals being in a communication need handy rhetorical tokens that are culturally comprehensible, only if their symbolic status has been previously preserved in language and is still widely recognized by members of a given culture through the language they use; *Marry Poppins* for the British culture and *Gumby* for the American culture serve as perfect examples of *compléments cognitifs*, which the speakers set in one communication act either recognize or not. Like the general national cultures, the social subvariants of each of these generate sociolinguistically 'local' cultural-language symbols. Urban slang, which has been the lexical reservoir for Tonia Lee's *Macbeth* adaptation, is not different in that respect. 'Nope' (Act 4 Scene 1) and 'hustler' (Act 3 Scene 1) are slang vocabulary examples playing the function of cultural-language symbols which enrich the urban slang dictionary of a vast slang speech community. Such communicational feature of urban slang generally makes it a considerable rhetorically flexible linguistic foundation to let the target intralingual translation audience grasp the original essence perceived through their present-day socio-cultural reality.

Translation procedures exploited in the literary adaptation of *Macbeth* into African American urban slang

Literary translation, which falls into the segment of specialized artistic translation activity, is always of a highly idiosyncratic nature. Furthermore, such culture-language adaptation product carries the flavour of the translators' personal communication style. I am of the opinion that the stylistic identity of a literary text can either be authored by its genuine maker (the author) or by its remaker (inter-/intralingual remaker – the translator); literariness as such ranges from original to translation in binary oppositions (Laiho 2009: 116–17).

Hence, Shakespearean *Macbeth* literariness has been ontologically immersed into slanging communities of an urban African American ethnic minority of NYC, which is undeniable proof of the eco-translation utility as an intralingual communication mechanism. How did the adaptation maker achieve the assumed goal to modify original literariness into the target text expectations? She must have applied innovative on-text operation tools, i.e., translation techniques. There are various collections of translation procedures promoted by proficient scholars (i.e. Newmark [(1988) 2003]; Snell-Hornby [2006]) useful in literary translations. I have observed a few on-text operations that let Tonia Lee introduce urban slang lexical units (words or phrases) into the adapted body of the translated original.

What appears as the most typical translation approach in literary adaptations is coining the target text, hypothetically addressed to everybody. Tonia Lee had an alternative perspective on her translation activity. The author of an 'intralingual remake' of *Macbeth*, however, opted for recreating the original meaning into her translation by adapting particular dialogue turns into the target audience's urban slang communication practices, without any alteration in the stylistic intensity of the slangy marker 'lingering' in the target text. The urban slangs 'Cuz', 'Sista' and 'Pops' have their non-slangy flavoured equivalents of substantially stylistically neutral categorization. The same holds true of such phrases as 'word on the street' and 'he has lost it', which successfully replaced the original turns of perfect stylistically non-biased characteristics; these attract no spectacular readers' attention as either stylistically top-formal or stupendously informal. Moreover, this is what urban slang is about. It is natural, catchy and slightly frivolous to apply in on-record conversational circumstances. Table 8.1 lists a few examples of how Lee applied the equivalence technique in her adaptation (bold by M.G.).

It is natural that for some translation researchers the technique of equivalence yields similar translation effects to the technique of approximation. In my experience as a practising translator and a translation researcher, these are two specialized terms naming the same specialized translation activity. I am inclined to promote the former one as it is more recognized across the translation theory literature.

The second most extensively used adaptation technique was compensation. Here, the original text has been lexically manufactured to slightly intensify the reception effect when it comes to the explicature functions of certain pieces of text. Table 8.2 records a few bolded instances of compensation (bold by M.G.). Both slangy 'hustler' and 'kick some butt' refer to certain original turns, which do not flag such rhetorical informal intensity in the original as 'hustler' and 'kick some butt' do in the target text. With this adaptation procedure, Lee has

Table 8.1 The adaptation technique of equivalence

The Shakespearian original of *Macbeth* Shakespeare (1993)	Tonia Lee's intralingual adaptation of *Macbeth* Lee (2008)
Act 1, scene 2 DUNCAN O valiant **cousin**! worthy gentleman! (p. 3)	Duncan: Word. That's good news. Way to go, **Cuz**! (Referring to Macbeth) (p. 2)
Act 1, Scene 3 FIRST WITCH Where hast thou been, sister? (p. 4)	First Witch: Yo, where have you been, **Sista**? (p. 3)
Act 2, Scene 2 LADY MACBETH Alack, I am afraid they have awaked, And 'tis not done. The attempt and not the deed Confounds us. Hark! I laid their daggers ready; He could not miss 'em. Had he not resembled My **father** as he slept, I had done't. (p. 19)	Lady Macbeth: Oh, shoot. Someone woke up. I left daggers ready for Macbeth in case he needed them. He can't miss them. Oh, man, I hope Macbeth doesn't screw things up. I would have killed Duncan myself if he didn't look like my own **Pops** when he was sleeping. (p. 15)
Act 5, Scene 2 CAITHNESS Great Dunsinane he strongly fortifies: **Some say he's mad**; others that lesser hate him Do call it valiant fury: but, for certain, **He cannot buckle his distemper'd cause**. (p. 60)	Caithness: **Word on the street is** that he has fortified the castle at Dunsinane. Some say he's crazy; others say he is just full of himself. I know one thing; **he has lost it! Big time**! (p. 49)

reinforced the informality level of a wide collection of original lexical elements. Stylistically, of course, the target text becomes slangy, which has been expected to happen and appreciated by urban slang users, keeping in mind the primary sociological motivation for using slang, which is the need for socialization (Widawski 2015: 99–116).

By the same token, I explain Lee's decision to introduce into her adaptation slangy words/phrases which mirror no contextual embedding traceable in the original text. Table 8.3 records the examples of enrichment (bold by M.G.) as the third most commonly used adaptation technique by Lee I have located in her text. The slang word stock enrichment adaptation technique is paradoxically highly functional when the translator aims to coin a concise variant of the original turn into another language. Indeed, it is so when it comes to a comparative textual analysis of both works; those target text dialogue turns adapted with at least

Table 8.2 The adaptation technique of compensation

The Shakespearian original of *Macbeth* Shakespeare (1993)	Tonia Lee's intralingual adaptation of *Macbeth* Lee (2008)
Act 3, Scene 1 MACBETH. So is he mine; and in such bloody distance, That every minute of his being thrusts Against my 'near'st of life: and though I could With barefaced power sweep him from my sight And bid my will avouch it, yet I must not, For **certain friends** that are both his and mine, Whose loves I may not drop, but wail his fall Who I myself struck down; and thence it is, That I to your assistance do make love, Masking the business from the common eye For sundry weighty reasons. (p. 32)	Macbeth: You will. Many people **put you types down**, but the world needs all kinds of people for it to run smoothly – the ganstas and **hustlers**, too. (p. 26)
Act 3, Scene 5 HECATE. Have I not reason, beldams as you are, **Saucy and overbold**? How did you dare **to trade and traffic with Macbeth In riddles and affairs of death.** And, I, the mistress of your charms, The close contriver of all harms (…) (p. 40)	First Witch: What's good, Hecate? You look mad tight. Hecate: I'm **pissed off** at you **smelly scumbags**! How come y'all didn't let me in on this whole Macbeth thing? I'm supposed to be your leader. Nothing like that should go down without me being involved. First Witch: My bad, Hecate. Second and Third Witches: Me too, **my bad**. (p. 33)
Act 4, Scene 3 MALCOLM. This tune goes manly. Come, go we to the King; our power is ready, our lack is nothing but our leave. Macbeth is ripe for shaking, and the powers above put on their instruments. **Receive what cheer you may, The night is long that never finds the day.** (p. 57)	Malcolm: That's what I'm talking about. I think our army is ready. Let's go **kick some butt!** (p. 45)
Act 5, Scene 3 MACBETH. ….. **Seyton-I am sick at heart, When I behold – Seyton, I say!** – This push Will cheer me ever or disseat me now. (p. 62)	Macbeth: (Calling) Seyton! Seyton! **Get your butt in here!** (Talking aloud) This war will either be my beginning or my end. (…). (p. 51)

Table 8.3 The adaptation technique of enrichment

The Shakespearian original of *Macbeth* Shakespeare (1993)	Tonia Lee's intralingual adaptation of *Macbeth* Lee (2008)
Act 1, Scene 7 **MACBETH** Prithee, peace: I dare do all that may become a man; Who dares do more is none. **LADY MACBETH** What beast was't, then, That made you break this enterprise to me? When you durst do it, then you were a man; And, to be more than what you were, you would Be so much more the man. Nor time nor place Did then adhere, and yet you would make both: They have made themselves, and that their fitness now Does unmake you. I have given suck, and know How tender 'tis to love the babe that milks me: I would, while it was smiling in my face, Have pluck'd my nipple from his boneless gums, And dash'd the brains out, had I so sworn as you Have done to this. (p. 15)	Macbeth: Chill! You are not that gansta! Those are my kids you are talking about! Watch **your mouth, woman**! Lady Macbeth: **Talk to the hand**. You don't know how to get anywhere in life. (p. 10)

a minor application of enrichment translation technique are noticeably more concise and include shorter sentences. Slang expressions are concise in form and compact with the semantic charge they carry, and this was probably the reason behind Lee's choice to apply that adaptation method.

The Shakespearian classics will always have fervent readers and fans. However, young children are rarely interested in literary pieces written with demanding vocabulary and complicated style. To introduce today's young audience to Shakespearian dramas and to sensitize new generations to their exceptional aesthetic value, translators have to refresh the surface layer of those texts. Lee's successful endeavours to implement a 'style shift' (Bucholtz 2011: 138) in *Macbeth* gave magnificent sociolinguistic results. A socially hermetic type of literary piece began to be accessible to the 'common folk' and its reception among young urban slang users has been cheerfully acknowledged, as I have learned, reading countless English language posts on various internet fora devoted to the social and cultural issues of the African American diaspora in the United States. Knowing this, I have decided to focus on qualitative dissimilarities between the two versions of *Macbeth* that can be proven in a dedicated stylometric study (see Olejniczak 2018).

The log-likelihood keyness corpus analysis of the two renderings of *Macbeth*

The following segment of the study presents only a fraction of qualitative corpus analysis which can be carried out on these two intralingual renderings of *Macbeth*.[1] The general point of reference in the stylometric analysis of the quantitative data incorporated for the study was a keyness test performed with the use of the log-likelihood test. This one has been grounded on the assumption that keyness above 6.6 is statistically significant (meaning that the differences between the corpora for selected items are not present due to chance alone). In addition, it accepts the thresholds proposed by Gabrielatos (2018) formulated on the basis of the following criteria: keyness of 13.8 proves very positive evidence that the contrast is present between the two corpora, keyness of 18.81 indicates that the evidence for differences is strong and keyness of 22.22 and above proves that the evidence for differences is very strong (Olejniczak 2018; 2022). The log-likelihood keyness corpus analysis appears to be a handy statistical tool for gathering some data, which would enable the translation researcher to draw demonstrable conclusions about how much the literary artefact has been modified within the act of artistic adaptation into another language/language variant.

This section outlines a few key grammatical differences between the two versions of *Macbeth* locatable thanks to the application of the log-likelihood keyness corpus text analysis. The keyness threshold above 6.6 in each of the following case analysis points is statistically significant. It ultimately signifies that the stylistic value of the original has been altered for the target text users' needs (see Skopos theory by Reiss and Vermeer [1984]).

Gerunds and continuous verbs, and conjuncts

The likelihood keyness test indicates noteworthy differences in the count of gerunds and present continuous verbs (Table 8.4).

Both versions of *Macbeth* contain gerunds to a very similar extent. Here, however, the continuous verbs appear in a higher number in Lee's adaptation (keyness of 142.6), which is solid proof of the substantial remodelling of the stylistic character of Lee's adaptation. At the same time, a considerable statistical difference between those two texts is observable for coordinating conjuncts

Table 8.4 The use of gerunds and continuous verbs

Item	Count ORIGINAL	Count SLANG	Keyness
Gerunds and continuous verbs VVG	114	320	142.6

Table 8.5 The use of conjuncts

Item	Count ORIGINAL	Count SLANG	Keyness
Conjuncts	780	378	79.3
And	566	245	79.2
But	124	67	9.2
Yet	56	12	23.5

(Table 8.5). This difference also impacts the stylistic manner in which the entire storytelling is to be conceptualized in a literary text.

'Yet', which has always been recognized as highly formal in use, is very scarcely used in Lee's adaptation. However, 'and' and 'but' in syntactic constructions are too stylistically predominantly recognized as formal. Their application triggers longer and far more complex sentences typical of the Shakespearian original.

Contractions

The difference in the use of contractions (e.g. 'he's,' 'they're,' 'I'm') between the two analysed texts is more than striking. The original *Macbeth* records just five instances of contractions, whereas Lee's adaptation holds the total number of 159 (with an average keyness of 73.7 for all three sections) (Table 8.6).

The use of contractions is typical of low registers, including slang, and urban slang is no different here. The first-of-all eco-translation strategy is the Schleiermacher's ([1813] 1973) 'domestication'. Vividness and naturalness mixed up with an egalitarian extravagancy of the street speech constitute today's urban slang. Contractions are one out of many cells forming the DNA structure of slang that is a perfect vehicle for adaptation through domestication, which Lee has achieved in *Macbeth*. This is another spectrum of corpora contrastive analysis proving that Lee's eco-stylistic approach to the literary adaptation of *Macbeth* was successful on the very pragma-rhetorical level; if Lee's target language

Table 8.6 The use of contractions

Item	Count ORIGINAL	Count SLANG	Keyness
First-person pronoun contractions (e.g., I'm)	0	66	103.5
Third-person pronoun contractions (e.g., he's)	5	56	59.6
Contractions of 'not'	0	37	58.2

localization focus was supposed to be locked on the African American urban slang users, the language conversational maxims applied by her in this adaptation must have resulted in simplifying grammatical constructions to the practical minimum. This happened, so it may be stated that Lee has achieved that goal.

Adjective Distribution

Another evidence of the compactness of Lee's adaptation of *Macbeth* is the quantitative distribution of adjectives (see Table 8.7). The degree of the figurativeness of the deep structure of every text message is induced by and grows accordingly along with the increase in the number of adjectives in the very surface layer of the text itself. Eloquent and passionate speakers/writers are industrious with adjectives turning sentences they use into a vivid tornado expressivity, which is the root cause of figurativeness as a rhetorical figure of conversation.

The contrastive qualitative analysis of both texts reveals a radical drop in the use of adjectives in Lee's adaptation in comparison to the original by more than one-third (35.6 per cent). Slang does need metaphors to thrive socially and spread among its users in the rhetorical spectrum of interpersonal communication. Nevertheless, slang shapes the extralinguistic communication sphere primarily with literal (unambiguous) references to its entities and artefacts (people, objects, events, etc.) we encounter daily. This is why in her adaptation work, Lee hammers the glow of the conceptual metaphors of the original. The aesthetic sensation of the stylistic rendering induced the readers of the original

Table 8.7 The use of adjectives

Item	Count ORIGINAL	Count SLANG	Keyness
Adjective	1063	658	37

to visualize the figurativeness of the source semantic domain shining through longer passages, dexterously coined adjective phrases, lexical doublets or archaic adverbs (e.g. 'therein') exploited today principally in specialized varieties of language (e.g. legal discourse).

Macbeth in urban slang yields fewer traces of Shakespearean metaphorical schemes for emotional states or conditions usually aroused by humans as social animals. The socio-cultural language communication patterns and traditions of people living in the time of Shakespeare, on the one hand, are shared with the ones used by us, living at the beginning of the twenty-first century. However, on the other hand, they are obviously dissimilar. Language economy leading to the melioration of language rules lowers the standards of language use itself, and the reduction of the use of adjectives in language contacts reduces the circulation of metaphors among individuals living in the social organism and sharing common cognitive communication space. This way, some Shakespearean *compléments cognitifs* may no longer be widely comprehended by the readers of Tonia Lee's adaptation as cultural-language symbols in today's universe of Anglo-Saxon communication.

Conclusion

The analytic section of the chapter has unveiled only a sector of the collected and researched data and a fragment of a longitudinal process of the intralingual translation investigation of similarities and differences between the two versions of *Macbeth* I have been involved in for the past years. Nevertheless, this insight has provided enough data to answer two research questions I formulated in the introductory part of this chapter.

- What translation tools – i.e., translation techniques – has Tonia Lee applied to her professional translation endeavour to achieve the African American urban street speech flavour in her adaptation of *Macbeth*?

Lee has manufactured her rendering of Macbeth with the use of equivalence, compensation and enrichment adaptation techniques to introduce (via enrichment) or induce (via equivalence, compensation) the African American slangy flavour into the stylistic environment of the street speech reflections one encounters in her adaptation written in urban slang. However, the word-for-word, faithful or descriptive translation techniques would eventually derail Lee's principal strategic aim to 'socially ecologize' the text, as I understand that if she had applied them in this adaptation challenge.

- What are the stylometric differences between the two renderings of *Macbeth*, basing the research on a qualitative corpus analysis?

Lee's adaptation of *Macbeth* is more concise, which is manifested by a relatively low number of conjuncts used. The stylistic level of the text itself is reduced to the casual (the use of gerunds and continuous verbs) and colloquial (an outstandingly high number of contractions present in Lee's adaptation) registers. The two analysed texts perhaps do not represent profoundly diverse renderings on all levels of linguistic structures. Nevertheless, the adaptation of *Macbeth* into urban slang is, beyond a doubt, a model example of an eco-translation performance.

Tonia Lee has efficaciously reduced the threat of intracultural incommensurability between Shakespeare and his African American urban slang users. This target sociolectal variety of American English adaptation is a complete literary artefact because Tonia Lee, being herself an African American teacher and social educator, knew exactly what necessary connotations and denotations urban slang lexical units had to evoke to let the target version readers 'capture' the essence of *Macbeth*. The eco-translation example of her authorship is founded on all stable sociolinguistic concepts of communication norms and systems in today's slang; however, a literary, cultural turn is achievable only through the act of linguistic domestication triggered by similar communication patterns:

> Young, urban, everyday members of the working class were some of Shakespeare's biggest fans. Many of the words in Shakespeare's plays were considered slang for its time. Therefore, a translation of *Macbeth* into urban street slang upholds the true spirit of Shakespeare's plays, as it reaches out to the same audience. (Lee 2008: iv)

Note

1 I would like to take this opportunity to thank Jędrzej Olejniczak, a doctoral student of the Institute of English Studies, University of Wrocław, Poland, who shared his knowledge and corpus analysis skills with me. This helped me arrive at the key conclusions on the stylometric status of certain details and dependencies between those two compared texts.

References

Abrahams, R. D. (1967), *Talking Black*, Rowley, MA: Newbury House.
Bauer, L. (2002), *An Introduction to International Varieties of English*, Edinburgh: Edinburgh University Press.
Baugh, J. (1983), *Black Street Speech: Its History, Structure, and Survival*, Austin, TX: University of Texas Press.
Bo, T. (2014), 'A study on advertisement translation based on the theory of eco-translatology', *Journal of Language Teaching and Research*, 5 (3): 708–13.
Bucholtz, M. (2011), *White Kids: Language, Race, and Styles of Youth Identity*, Cambridge: Cambridge University Press.
Cukor-Avila, P. and G. Bailey (1996), 'The spread of urban AAVE: A case study', in J. Arnold, R. Blake, B. Davidson, S. Schwenter and J. Solomon (eds), *Sociolinguistic Variation: Data, Theory, and Analysis*, 469–85. Stanford: CSLI.
Gabrielatos, C. (2018), 'Keyness analysis: Nature, metrics and techniques', in C. Taylor and A. Marchi (eds), *Corpus Approaches to Discourse: A Critical Review*, London; New York: Routledge. https://repository.edgehill.ac.uk/9454/.
Garcarz, M. (2013), *African American Hip Hop Slang: A Sociolinguistic Study of Street Speech*, Wrocław: Atut.
Green, J. (2010), *Green's Dictionary of Slang*, London: Cambers Harrap Publishers Ltd.
Green, L. J. (2002), *African American English: A Linguistic Introduction*, New York: Cambridge University Press.
Guo, Y. and Y. Wan (2022), 'Retracing the history of "Word for Word", "Sense for Sense" translation – confronting and inheriting of the ancient roman translation theories', *Open Journal of Modern Linguistics*, 12 (5): 568–77.
Jakobson, R. (1959), 'On linguistic aspects of translation', in R. A. Brower (ed.), *On Translation*, 232–40. Cambridge, MA: Harvard University Press.
Labov, W. (1972), *Language in the Inner City: Studies in the Black English Vernacular*, Philadelphia, PA: University of Pennsylvania Press.
Laiho, L. (2009), 'A literary work – translation and original: A conceptual analysis within the philosophy of art and translation studies', in Y. Gambier and L. van Doorslaer (eds), *The Metalanguage of Translation*, 105–22. Amsterdam/Philadelphia: John Benjamins Publishing Company.
Lee, T. (2008), *Shakespeare in Urban Slang*, New York: Urban Youth Press.
Newmark, P. ([1988] 2003), *A Textbook of Translation*, Harlow: Pearson Education.
Olejniczak, J. (2018), 'Using corpora to aid qualitative text analysis', *The Journal of Education, Culture, and Society*, 9 (2): 154–64.
Olejniczak, J. (2022), *Private Academic Consultation on Corpora Qualitative Text Analysis*, Wrocław: University of Wrocław, Poland.
Reiss, K. and H.-J. Vermeer (1984), *Grundlegung einer allgemeinen Translationstheorie (Linguistische Arbeiten 147)*, Tübingen: Max Niemeyer Verlag.
Rickford, J. R. (1998), 'The Creole origins of African-American vernacular English: Evidence from copula absence', in S. Mufwene, J. R. Rickford, G. Bailey and John

Baugh (eds), *African American English: Structure, History, Use*, 154–200. London/ New York: Routledge.

Schleiermacher, F. ([1813] 1973), 'Über die verschiedenen Methoden des Übersetzens', in H. J. Störig (ed.), *Das Problem des Übersetzens*, 38–70. Darmstadt: Wiss. Buchgesellschaft.

Shakespeare, W. (1993), 'Macbeth', in *The Complete works of William Shakespeare*, San Francisco: World Library, Inc.

Smitherman, G. (1977), *Talkin' and Testifyin': The Language of Black America*, Boston, MA: Houghton Mifflin Company Boston.

Smitherman, G. (1998), 'Word from the hood. The lexicon of African American vernacular English', in S. Mufwene, J. R. Rickford, G. Bailey and John Baugh (eds), *African American English: Structure, History, Use*, 203–25. London/New York: Routledge.

Snell-Hornby, M. (2006), *The Turns of Translation Studies*, Amsterdam/Philadelphia: John Benjamins Publishing Company.

Venuti, L. (1995), *The Translator's Invisibility: A History of Translation*, London/New York: Routledge.

Widawski, M. (2015), *African American Slang: A Linguistic Description*, Cambridge: Cambridge University Press.

Fictional Ekphrasis Representing Childhood Trauma in M. Atwood's *Cat's Eye*

Polina Gavin

Introduction

Cat's Eye is a 1988 novel written by Canadian author Margaret Atwood. *Cat's Eye* is a fictional autobiography narrated by Elaine Risley, an artist who comes back to her hometown of Toronto to attend a retrospective exhibition of her work. When visiting Toronto, Elaine is overwhelmed by the traumatic memories of a power-driven relationship with her former friend Cordelia. Being brutally bullied as a child by a group of girls, abetted by Cordelia, Elaine develops post-traumatic stress disorder (PTSD), majorly manifested in dissociative amnesia. The novel follows the gradual return of Elaine's emotive memories and consequent recognition of the trauma that allows for the re-discovery and restoration of the character's coherent and wholesome self.

Atwood (1986) discusses the unconventionality of 'the best girlfriend genre' that acquires a darker tone, where the violation of friendship and betrayal is more unexpected and thus 'all the more unbearable'. The author highlights that female relations are complex and power-driven even in childhood: 'little girls are not made of sugar and spice and everything nice' (BBC Bookclub 1999). On several occasions, Atwood admits that she has received a multitude of responses from women who could identify with Elaine's victimization and confessed to being bullied mentally and physically by their girlfriends to a much greater extent than described in the novel (Terkel 1989; BBC Bookclub 1999). The high level of the readership empathetic identification with the protagonist allows Atwood to contend that Cordelia is more of a prototypical phenomenon in women's friendships: 'we probably all do have a Cordelia' who is 'indelibly engraved' in our memories (Terkel 1989).

Literary scholars approach the portrayal of power imbalance and trauma in *Cat's Eye* from different angles. The scholars refer to art as a subconscious representation of trauma and provide a psychoanalytical rationale for its role in Elaine's identity construction (Banerjee 1990; Dvořák 2001; Vickroy 2005). While, at first, the mediated trauma remains unrecognized, an artistic process gradually facilitates the return of agency and sense of control to the protagonist. This explains Elaine's propensity for art (Vickroy 2005) and the salient role of paintings in representing the protagonist's PTSD. Despite the abundance of work on trauma in *Cat's Eye*, the research touches lightly on how the paintings mediate a traumatized mind in the novel. An exception to this is a recent study focusing on the literary representations of Elaine's and Cordelia's (self)-portraits (Paiva 2022). Paiva (2022) attends to fictional artworks as visual metaphors that voice the underlying themes of the novel and allow the protagonist to question, integrate, and restore the wholeness of their distorted and fragmented self.

The existing research warrants further exploration of Elaine's fragmented identity through the linguistic representation of her art. This chapter examines how fictional ekphrasis mediates the protagonist's mind style, a phenomenon that identifies idiosyncratic ways of portraying the character's world of view in literary fiction (Fowler 1977). The chapter draws on cognitive poetic scholarship in ekphrasis (Verdonk 2005; Gavins 2012; Panagiotidou 2016; 2017; 2022a; 2022b) and aims to expand the current knowledge of fictional ekphrasis as a linguistic representation of an artwork that does not have a physical referent outside the dimensions of a literary text. In my analysis, I employ Text World Theory (TWT henceforth) and Cognitive Grammar (CG henceforth) that allow for a nuanced examination of the character's worldview encoded in the representations of imaginary art.

The chapter is organized in the following way. The next section provides an overview of cognitive poetic research on ekphrasis and sketches the methodological tools used for its linguistic analysis. It is followed by the section that focuses on the concept of mind style and elucidates the ways the protagonist's cognitive idiosyncrasies stemming from PTSD are presented in the text. The next section traces the build-up of the protagonist's terrorized mindset, where the negation of self, dissociation, and repression of memory are indicated in language. The final section examines two cases of fictional still lifes, with each of them approached from the distinct versions of the protagonist's mind in the course and aftermath of a traumatic experience. The analysed instances of fictional ekphrasis represent Elaine at various stages of dealing with trauma and follow her journey to regaining her memories and reconciling with the past.

Fictional ekphrasis

Ekphrasis is a literary phenomenon commonly defined as a retelling of a visual artwork. Rooted in ancient Greek rhetoric, ekphrasis served as an exercise that aimed to evoke a represented subject vividly before the eyes of the audience (Webb 2009). Ekphrasis held the potential to transport the audience into the imaginative world of a narrative, thus amplifying the listeners' emotional involvement with the story and increasing the persuasive power of the message. Since then, an ekphrastic tradition has significantly transformed, leading to ekphrasis addressing a cross-media relationship between arts. The indispensable experientiality of ekphrasis and its connection to wider sociocultural and personal knowledge have triggered scholarly interest in cognitive stylistic studies. The existing stylistic discussions concern the cross-media relationship between artefacts and cognitive implications they have for the reader's processing, as well as interpretative effects that emerge when reading an ekphrastic text (Verdonk 2005; Gavins 2012).

A comprehensive account of poetic ekphrasis as a cognitive phenomenon is presented by Panagiotidou (2016; 2017; 2022a; 2022b). An ekphrastic experience is co-created between the reader and the implied personae of an artist and an author who contribute to constructing an ekphrastic response, even being displaced in space and time. Panagiotidou (2016; 2017) showcases a holistic construction of a cross-media text-world in the reader's mind by tracing parallels in the visual arrangements of the painting and the figure-ground text setting. She highlights four features which may be present in ekphrastic texts up to varying degrees: representation, narrativization, transposition, and collaboration. When combined, the defined qualities manifest ekphrasticity in a text – an ability to evoke 'a cognitive response to a text that (re-)presents, narrativizes, alludes to, or evokes a work of art' (Panagiotidou 2022b: 48).

The discussed case studies focus on 'actual ekphrasis' (Hollander 1988: 209), where the referenced artwork is real and identifiable. Fictional works of art have not yet received much scholarly attention in the cognitive stylistic field, with one exception being Panagiotidou's (2022a) ecostylistic analysis of the narrating persona's conceptualization of the natural environment portrayed in an imaginary painting. Panagiotidou (2022a) demonstrates how ekphrastic writing becomes a channel for the expression of the narrator's idiosyncratic worldview. This chapter develops the discussion of ekphrasis confined within a literary text, where artwork is conceived and observed by literary characters or, in this case, the changing mind states of the protagonist.

I use Mason's (2019) theory of narrative interrelation that explains how intratextual relations are initially made in the mind but can be linguistically traced. The reader's interpretation of intratextual links emulates the ways we make sense of real-life experiences and interweave them into stories. Mason's (2019) framework provides insight into how intratextual interrelations can create cohesion, advance plot, and introduce certain stylistic effects. An ekphrastic retelling of an imaginary artwork cannot rely on a visual referent from the outside world; therefore, the context and the meaning of a visual scene need to be fleshed out intratextually. Fictional ekphrasis in *Cat's Eye* is built across the narratives of past and present, which reflect Elaine's dynamic characterization during her experience, recognition, and recovery from trauma (Banerjee 1990). The description of an artwork and the context that inspired its creation feature different manifestations of gaze – a focalized perspective of Elaine as a severely bullied child and Elaine as an adult with the fragmented memory as a PTSD after-effect. I argue that the changing perception of fictional art represents different versions of Elaine's mind style. I will further outline the methodological premise of TWT and CG that I use to sketch more broadly how a trauma-affected mind cognitively and linguistically refracts in ekphrasis across the narrative layers.

TWT is a 'cognitive discourse grammar' (Giovanelli 2013: 11) founded on the premise that cognitive processing of discourse involves the construction of a text-world – a representation of the discussed situation in the mind. The construction of a text-world is shaped by the 'principle of text-drivenness' that underpins the contextual information required for fleshing out a representation (Werth 1999: 103). A text-world consists of 'world-builders' that define the setting and 'function-advancers' that move the narrative forward (Gavins 2007: 56). The dynamic change of text-worlds is marked by 'world-switches' (Gavins 2007: 74). Participants and characters project their distinct versions of 'enactors', that are consistent with the text-world setting, onto a new text-world (Gavins 2007: 40). The world-switch is triggered by shifts in time, location, and perspective. The new text-worlds are created by metaphorical mappings, negation, and hypothetical constructions promoting alternative plot developments.

CG proposes an account of language structure that is inherently rooted in the embodied nature of language. The CG central idea is a 'construal' that encodes 'our manifest ability to conceive and portray the same situation in alternate ways' (Langacker 2008: 43). The CG central premise lies in how the conceptual content may be presented, or construed, in different ways that are expressed through meaningful grammatical configurations. Construal is a linguistic arrangement for how the scene is conceptualized by the writer and viewed by the reader. The four construal phenomena are specificity, focusing, prominence, and perspective (Langacker 2008: 55–89). Specificity refers to the level of detail that the conceptual content features; focusing arranges the selected content into foreground and background with a maximal or immediate scope of coverage; prominence indicates how a particular part can be profiled against its conceptual base; and perspective demonstrates how the position of the narrator/reader is grounded in the construed scene. The involvement of the narrator/character can vary, thus reflecting a subjective (emphasis on the conceptualizer) or objective (emphasis on the content) conceptualization of the scene.

Elaine's mind style affected by PTSD

The concept of mind style in cognitive stylistics covers a body of research that examines how fictional and non-fictional minds are represented and captured by readers in language. Mind style is 'any distinctive linguistic representation of an individual mental self' (Fowler 1977: 103) and is most prominent in the studies of characters who display an unconventional worldview due to some cognitive impairment or an extreme effect of the surrounding physical and social environment. In this sense, mind style falls into its strictest category as 'the representation of a highly deviant or at least unusual worldview' (Stockwell 2009b: 125). The inclination of stylisticians to examine the deviancy of cognitively impaired or criminally dispositioned minds is justified by the propensity of such mind styles to hold higher interpretative significance (Leech and Short 2007). The ongoing research encompasses a range of minds that differ in the orthodoxy of their 'mental set' and positions them along the cline, with extremes of the 'normal' and unconditional conceptions of the world placed at its ends (Leech and Short 2007: 151).

Stockwell (2009b: 124) introduces a 'cline of viewpoints' with idiosyncratic worldviews given priority to attribute a mind style. Nuttall (2018: 80–1) reconfigures the model into a 'cline of construals' to highlight how the range of viewpoints can be psychologically and socially motivated and then enacted via the CG machinery. The latest advancement to the model is proposed by Giovanelli (2022), where he argues in favour of Leech and Short's continuity of mind styles. From a CG perspective, all construals have the capacity to evoke a mind style that can be conventional to varying degrees. The major advantage of such approach resides in capturing a continuum of different attitudes and minds.

Panagiotidou (2022a) demonstrates how a mind style that does not display an overt cognitive impairment is elucidated through ekphrastic writing. An 'uneasy narrator' (Panagiotidou 2022a: 19) expresses a peculiar worldview of discontent and entrapment when observing a fictional work of art.

A range of studies explores the mind style of characters displaying memory impairments as an after-effect of a severely traumatizing experience or a persisting symptom of a cognitive condition (for a good overview, see Giovanelli 2022: 87–92). When experiencing an overwhelming event, the mind becomes unable to process it at the moment of occurrence and suppresses the memory as a coping strategy. The blocked memories reside in the subconscious and haunt the victim until they are resolved (Kurtz 2018). The saliency of memory fragmentation in PTSD is featured in the continuous, unexpected, and uncontrollable intrusiveness of the past in the present moment (Diedrich 2018), leading scholars in literary trauma studies to label PTSD a 'disorder of memory' (Leys 2000: 2).

In *Cat's Eye*, Elaine is presented as extremely unreliable due to severe memory loss. The evasiveness of Elaine's memories is manifested in an unconditional narratological presentation that is common to autobiographical accounts of characters with cognitively augmented minds (Emmott and Alexander 2015). The novel is structured as a retrospective in a twofold sense: each chapter commences with the present narrative of Elaine's visit to Toronto and then transitions to a particular memory of her past narrated by a much younger enactor of Elaine. The non-linear narrative is foreshadowed at the start of the novel with Elaine's reflections on time using the conceptual metaphor TIME IS SPACE, where time is an obscurely layered dimension with memories haphazardly surfacing the mind. The plot fragmentation is balanced out by the chronological progression of events within the storylines of past and present. The storylines are spread out across the novel as continuous text-worlds grounded in their distinct spatiotemporal dimensions. The storylines stretch and compress time – the present Elaine's enactor goes for a short work trip to oversee an art retrospective, while the past Elaine's enactor lives through '[her] life entire' (Atwood 1990 [1988]: 468) from the age of eight to her mother's death several years before the art exhibition. The narrative loops backward are not deemed as flashbacks but given a more significant temporal depth of 'narrative braids' (Banerjee 1990: 515) or 'narrative arcs' (Harrison 2017: 138). The alternate plot structure of a 'circular return' emulates the wave-like disentanglement of memory and defies the traditional linear frame of a künstlerroman that *Cat's Eye* is commonly ascribed to (Osborne 1994: 95–6).

Elaine's traumatic experiences are externalized through painting, when a creative task serves as an attempt to explicate trauma in visual images rather than language. The fictional paintings transcend visuality of 'harrowing mental images' of a traumatic experience (Armstrong and Langås 2020: 5). The role of ekphrasis as a 'literary mediation for the act of witnessing' (Armstrong and Langås 2020: 5) becomes more complex in this case because it mirrors the psychological state of a trauma survivor, while, at the same time, the conveyed visual scene is unable to represent an experience directly: thematically or stylistically (Luckhurst 2008). Elaine's paintings reflect the fragmented perception of self with the changing enactors of the protagonist, who experiences, represses, and recognizes trauma. The manifestation of gaze is particularly salient in the art that elicits trauma. Panagiotidou (2022b) considers gaze a powerful interpretative mechanism as it projects the origo – a 'zero reference point of subjectivity' onto a text-world (Gavins 2007: 36). The origo encompasses an embodied cognition that appertains to a certain moment in the narrative, and therefore holds the potential to attribute a mind style. In *Cat's Eye*, a child-enactor of Elaine mediates an experience of being bullied that will have consequently inspired an artwork, while an adult-enactor of Elaine observes her artistic practice in detachment and confusion from what her art is aimed to portray. I will further linguistically trace how the bullying reaches a critical point that triggers Elaine's split from memory and self.

Burying and the negation of self

Storytelling is a universal human propensity to organize experiences into coherent personal narratives 'for their comprehension, memorization and recall' (Lambrou 2014: 33). The recollection of an individual's life story conduces to the reconstruction of their identity, which can be traced through the stylistic choices (Lambrou 2014). The storytelling of trauma experience and survival is the protagonist's attempt to make sense of the occurred events. 'Until we moved to Toronto I was happy.' – an enactor of a younger Elaine starts narrating the past chronicles with a powerful juxtaposition of her family's nomadic lifestyle in the Canadian bush with their decision to settle down in Toronto. The temporal 'until' is a borderline that presumes two distinct life styles, two separate worlds, and hence two attitudes that significantly foreshadow the upcoming future, while leaving it unexplained. The reader then is taken back to the child-enactor of Elaine reliving her struggles to fit in at school. Having spent early years in the sole company of her brother, Elaine has to conform to the social behaviours customary to girls, such as pretence and the non-spoken shame for female physicality. Elaine constantly feels awkward, embarrassed, and self-conscious, which makes her highly susceptible to the limitless influence of her so-called friends. The girls intimidate and terrorize Elaine as an outsider who needs to be improved to match the accepted standards in the society. Elaine's compliance is voiced by the deontic modality of restriction and obligation that re-occurs in many other parts of her storytelling, conveying the increase of subordination and loss of control. The epicentre of the girls' bullying happens when Elaine is 'playfully' buried alive on Halloween:

> I'm supposed to be Mary Queen of Scots, headless already. They pick me up by the underarms and the feet and lower into the hole. Then they arrange the boards over the top. The daylight disappears, and there's the sound of dirt hitting the boards, shoveful after shoveful. Inside the hole it's dim and cold and damp and smells like toad burrows.
>
> (Atwood [1988] 1990: 125)

The perspective from which the burying scene is construed can reveal an additional information on Elaine's perception of self. The role of Mary Queen of Scots is imposed on her ('I'm supposed to be'); the girls possess agency during the process of burying ('they pick me up', 'lower', 'they arrange'), while Elaine is assigned a passive role of a 'patient' or a 'mover' (Giovanelli and Harrison 2018: 88–9). The subjective construal of the scene is promoted by an implied understanding of Elaine being a conceptualizer of the event. A younger Elaine's enactor narrates the scene but, simultaneously, occupies an objective stance which conveys the ambivalence of her involvement. For example, 'me' places emphasis on the object of conception and underpins an objective construal. The objective presentation is further relayed by a 'then' clause used in sequences (Lambrou 2014).

Elaine's fluctuating involvement is amplified with orientational deixis ('pick me up', 'lower into the hole', 'boards over the top', 'inside the hole') that marks Elaine's bodily containment in the hole. The verba sentiendi assist in the modelling of Elaine's presence, where the reader co-construes the dimming light, hears the shovefuls of dirt hitting the boards, feels cold and dampness. The oscillation between the subjective and objective presentation of the event is highlighted via the grammatical placement of the conceptualizer offstage but the definite reference to the sound and light. The nominal grounding creates an assumption that the conceptualizer identifies a particular setting which contrasts the simultaneous objective conceptualization of the scene. Such fleeting shifts in the construal are salient in reflecting the changes in the focalizer's conceptual distance and presence (Tabakowska 2014; Panagiotidou 2022a). Elaine's matter-of-factual coverage of the burying is devoid of emotion, and thus conforms to the mode of a 'narrative report' (Fludernik 1996: 71), where the protagonist retells the narrated event as if by witnessing rather than directly participating in it. Similar findings have been reported by Lambrou (2014) in an examination of a survivor's retelling of the 7/7 terrorist attack. As the story is narrated from the first-person perspective and renders the protagonist's experiences, the linguistic evidence of Elaine's non-presence may signal the traumatic impact of an extremely frightening occurrence. The protagonist's dissociation from the traumatizing event is stylistically manifested in the ambivalent construal of (non)presence. Elaine's experiential presence in the body but not in the mind marks the starting point for Elaine's dissociation as a coping technique when dealing with trauma.

The burying incident escalates to the fragmentation of memory as a more severe repercussion of trauma (Herman 1992). In consequence, the reader evidences a drastic change in the protagonist's mind style, whose retrospective account of the event is rendered through the consciousness of an older Elaine's enactor. The reflecting adult Elaine is not identical to the present Elaine organizing a retrospective in Toronto. The recollection of the burying is narrated from the perspective of a particular version of Elaine, who is impossible to identify in space and time:

> I have no image of myself in the hole; only a black square filled with nothing, a square like a door. [...] The point at which I lost power. Was I crying when they took me out of the hole? It seems likely. On the other hand I doubt it. But I can't remember.
>
> (Atwood [1988] 1990: 126)

Although the quoted episodes follow one another in almost immediate succession, an older Elaine is now disoriented and confused. Elaine accepts that burying is real; however, she negates her bodily and mental presence at the event to the extent that she struggles to reconstrue the scene. In contrast to a younger enactor of Elaine, who sequentially scans the process of burying and strongly grounds her embodied presence, an older Elaine cannot re-orient herself in the surroundings. The narration becomes negatively shaded via words of estrangement (Simpson 1993), such as 'perhaps', 'it seems likely', rhetorical questions, epistemic and perception modality ('I doubt it', 'I can't remember'), which convey Elaine's unsuccessful interpretative efforts to restore the wiped-out memory. Elaine's reflexivity foregrounds the summative scanning of the experience: she realizes the traumatizing ramifications of the incident ('The point at which I lost power.'), allowing the reader to model for a mind that is afforded a full retrospection. However, the reader's mind-modelling here is biased: the reader conflates the information imparted to the mind of a younger Elaine with a summative account of the event's consequences. The minds of Elaine's enactors though remain distinct: Elaine in retrospect is dissociated and devoid of voice or agency, while the older Elaine's self is fragmented and entrapped between the realms of childhood and adulthood.

Still lifes and self-harm

I will further trace the readerly construction of two art pieces that are focalized through the consciousness of distinct enactors of Elaine and will demonstrate how linguistic presentation of art can shed light onto the protagonist's mind exacerbated by PTSD. An older reflecting enactor of Elaine takes the habit of painting objects that are non-present in her worldview at the moment of the narration. Elaine guesses that the painted objects are her memories but they arrive in isolation from any context:

> I paint a silver toaster, the old kind, with knobs and doors. One of the doors is partly open, revealing the red-hot grill within. […] I paint a wringer washing machine. […] The wringer itself is a disturbing flesh-tone pink.
> (Atwood [1988] 1990: 394)

The artist's subjective conceptualization is manifested in reference to the past: both the toaster ('the old kind') and the wringer washing machine are household items of the first half of the twentieth century, and Elaine acknowledges her familiarity with them. Except that, the stylistic presentation of still lifes reflects the artist's detachment. Elaine is more of an observer than a creator: the finite verb 'I paint' allows for the summative scanning of the event, foregrounding the resulting scene instead of the painting process. The world-building of the scene is extremely scarce and motionless. The pictorial elements of the objects are introduced with the existential 'to be' that places the conceptualizer offstage and promotes an efferent rather than aesthetic experience of the painting (Rosenblatt 1970). Elaine's inability to restore the memories that gave rise to her art is grammatically featured in varying nominal grounding. Grounding strategies refer to techniques that direct the attention to the referent and assign it with a certain degree of specificity (Giovanelli 2018).

Elaine's construal of a still life features overt indefinite grounding where the toaster and the washing machine, being mentioned for the first time, are used with an indefinite article. An indefinite article provides a larger set of possible referents that the reader can choose from, as opposed to a definite article that restricts the reference to one particular candidate (Langacker 2008). Giovanelli (2018) illustrates how indefinite grounding can convey the character's cognitive inability to reconstruct past events. He argues that the gradual shift of construal towards definite grounding represents the mind in the process of recalling a forgotten memory. Similar to Giovanelli's findings, Elaine's repressed traumatic experiences are grammatically mirrored in the use of indefinite articles. What stands out is an overt specificity of 'the red hot grill', where a definite article profiles one out of many instances of an object. The specificity suggests a certain level of recognizability of an instance that requires some shared understanding between the reader and the character (Giovanelli and Harrison 2018). The overt definite grounding functions as an 'attractor of definiteness' (Stockwell 2009a: 31) that is likely to capture the reader's attention.

The unconventional nominalization 'a pink' in the description of the washing machine is another deviant grammatical feature worth discussing. The metonymical transfer of colour onto the object in its entirety foregrounds the 'pinkness' of the wringer as its most salient and holistic feature. The evaluative 'disturbing' in conjunction with 'flesh-tone' profiles Elaine's ambivalent perception when it is unclear whether the foreshadowing of the ominous is an intention or a remnant of the repressed memory. In any case, the negative lexis is likely to act as an intratextual 'specific unmarked' reference (Mason 2019: 82–3) that navigates the reader's interrelation of two distinct layers conveying trauma experience and trauma repression.

The conceptualization of fictional art is only possible if the reader reconstrues the memory underlying an artistic impulse. It is worth mentioning that a claim for the fictional nature of Elaine's art can be questionable due to the notable autobiographical quality of *Cat's Eye*. The striking resemblances in Elaine's and the author's childhood have led the scholars to examine a conflation of fiction and autobiography in the novel (Cooke 1992; Givner 1992). Atwood herself urges readers to avoid such 'autobiographical fallacy' (Hubbard 1989: 206). She states in the disclaimer that Elaine's paintings do not exist, but their creation has been influenced by certain visual artists. The tangible influence of real-world art on Elaine's paintings may enrich the reader's experience of an artwork if they have a relevant narrative schema in their 'mental archive' (Mason 2019, 72). Otherwise, it is safe to say that Elaine's art is primarily fictional and needs to be fleshed out intratextually.

As illustrated previously, Elaine's mind style conveys the absence of a holistic conception of her life. The countless unidentifiable versions of Elaine narrate the forgotten events in the present, creating an impression of not a memory recall but reliving (Fisher 2015). 'It is the older Elaine who speaks, but the younger Elaine who focalizes' (Fisher 2015: 24), where the reader is given 'a record of things as the child saw, felt, understood them' (Rimmon-Kenan 2002: 75). The reader has privileged omniscient access to Elaine's feelings and thoughts that she herself struggles to puzzle together. The reader is 'pattern-hunting and consistency-building' (Brooks Bouson 1993: 168) for the details that ultimately recur in Elaine's art. The intratextual context for still lifes stems from a desperate time ensuing the burying incident when Elaine is driven to self-mutilation that escalates to suicidal impulses. Elaine seeks for a discreet way to avoid the girls' company and is reluctant to leave home as her perceived safe space. Elaine's uneasiness is expressed through the excessiveness of detail in the world-building of the kitchen. She describes a toaster:

> The toaster is on a silver heat pad. It has two doors, with a knob at the bottom of each, and a grid up the center that glows red-hot. […] I think about putting my finger in there, onto the red-hot grid.
>
> (Atwood [1988] 1990: 141)

The ekphrastic context explains the grammatical deviance of 'the red-hot grid' and fleshes out its poignancy as an emotionally charged representation of self-harm. The toaster is identical to the 'fictional' toaster in Elaine's still life. The comparable description of the toaster achieves high 'representativity' (Panagiotidou 2022b: 90) of the portrayed object. Given the reader's omniscience and ability to sequentially construct the narrative plot, they will be likely to identify a specific unmarked intratextual reference 'the red-hot grid' that will reiterate in the ekphrastic retelling later in the story. The prominence of the toaster grid is emphasized through repetition and the foregrounded deictic orientation ('in there, onto'). An epistemic modal-world 'I think of putting my finger in there ' continues an action chain of Elaine making a toast, where she first describes how she turns the toast over several times and then how it slides down. The action chain maps the routineness of making a toast onto Elaine's self-harming impulse and creates a sense that such thoughts are commonplace in her life.

The wiped memory of the wringer washing machine belongs to the relatively same period, when Elaine is eager to do house chores to avoid the girlfriends. Elaine's job is to rinse the wash, and she watches the washing machine working:

> I watch it, hands on the edge of the tub, chin on hands, my body dragging downward from this ledge, not thinking about anything. [...] I feel virtuous because of all the dirt that's coming out. It's as if I myself am doing this just by looking.
>
> (Atwood [1988] 1990: 144)

A high degree of specificity profiles a restricted viewing frame of Elaine's pose. The progressive form 'dragging' provides immediacy and highlights how Elaine's absorption into the scene grows to a bodily level. The negated epistemic modal-world 'not thinking about anything' creates an ambivalent impression, where, first, the contrasting text-world is clued with Elaine's thoughts being pre-occupied but then, consequently, negated. The ambiguous presentation of Elaine's mind seems to be 'disingenuous or deliberately evasive' (Fisher 2015: 25), conveying the sense of 'burying' the character's intentions behind the purported blankness of her thoughts. Elaine's immersion in the washing is further foregrounded by epistemic 'I feel virtuous' and hypothetical 'as if I myself am doing this' modal-worlds, which map the enactor's anxiety and desire to be seen as worthy onto the physical dimension where TO BE CLEAN IS TO BE BETTER. The metaphorical mapping is further unravelled in the wringing of the wash, when Elaine imagines her body being compressed by the wringers:

> [T]he arms or the shirts ballooning with trapped air, suds dripping from the cuffs. I've been told to be very careful when doing this: women can get their hands caught in wringers, and other parts of their bodies, such as hair. I think about what would happen to my hand if it did get caught: the blood and flesh squeezing up my arm like a traveling bulge, the hand coming out the other side flat as a glove, white as paper. This would hurt a lot at first, I know that. But there's something compelling about it. A whole person could go through the wringer and come out flat, neat, completed, like a flower pressed in a book.
>
> (Atwood [1988] 1990: 145)

The ambivalent 'the arms or the shirts' creates a complex layering of more than one conceptual base: while the reader is sure that these must be the clothes being washed, they construct a fleeting text-world with more gruesome undertones of a body caught between the wringers. The epistemic modal-world 'women can get' profiles a rich scope of body parts that the reader can infer from when conceptualizing the scene. The extension of the scope and its subsequent narrowing to a less prototypical variant of a body part ('such as hair') defeat the reader's expectations and make them adjust the inferred modal-world. The reporting of a speech act 'I've been told' infers that Elaine is aware of the potential danger, although she does not seem to be warned by it, but fascinated. She describes her flesh and blood being squeezed with peculiar specificity, where the progressive verb forms 'squeezing', 'coming out' make the experience more immediate and realistic.

The eerie sense of how real the projection pertains to be increases the reader's closeness to the scene. The recurrent profiling of a hand foregrounds the conceptual base of now Elaine's body being caught between the wringers, with 'the hand coming out' presented as an independent entity detached from the body. The rich imagery of the scene is conveyed with the same emotional detachment, where the concept of pain is not given much attention and left as an afterthought. The epistemic modal-world 'I know that' acknowledges Elaine's awareness of the possible consequences, but her fascination with the idea of self-abstersion surpasses the pain caused by self-harm. The epistemic modal-world ('a whole person could go through the wringer') launches the metaphor TO BE COMPRESSED IS TO BE COMPLETE. Elaine's divergent mind style is reflected in the idiosyncratic features mapped from a source domain to a target domain. In Elaine's mind, a physical transformation brings change to the self. The simile 'like a flower pressed in a book' maps the characteristics of 'flat' and 'neat' onto the body being flattened by the wringer, with the completeness added by Elaine as an aspired state of self and mind. The literal compression of the body would inevitably lead to a serious injury or even death; however, Elaine's troubled mind does not conceive an urge for purgation as an impulse for self-harm and potential suicide.

The conceptualization of a person compressed by the wringers from the point of view of Elaine's child-enactor is utterly objectifying and positive. The intratextual knowledge supplied by the forgotten memory enriches the reader's conceptualization of the stylistically marked pinkness of the washing wringer in the still life. The intratextual interrelation is sustained not by lexical repetition but by idiosyncratic metaphorical mappings that profile the schemas of trauma, self-harm, and suicide. The stylistic strategies used in ekphrastic descriptions and contexts showcase the differing mind style of Elaine's enactors, whose worldviews 'play off' each other and through contrast uncover cognitive habits of PTSD.

Conclusion

This chapter employs the stylistic frameworks of TWT and CG to demonstrate how fictional ekphrasis mediates mind style in the aftermath of childhood trauma. Previous studies in ekphrasis have largely focused on writing about existing and identifiable art. This case study develops the discussion of ekphrasis confined within a literary text where an artwork is observed by the changing mindsets of the protagonist. The reader finds an artwork description and its prompting context scattered across the narrative braids of the novel and is compelled to interrelate the disconnected fragments alternatively to the protagonist, who exhibits severe memory loss amongst other PTSD cognitive impairments and is unable to do so.

The close textual analysis of ekphrastic writing allows to trace the degrading dynamics of a trauma-affected mind. The complex interweaving of the Elaine's versions of self narrating the events and then struggling to remember them is marked by lexical and grammatical realizations of the protagonist's mind style. I demonstrate how the stylistic manipulations with nominal grounding, unconventional nominalization, and negative lexis in the descriptions of art relay an unreliable restricted perspective of an older Elaine, which, in turn, explains her inability to identify the depicted objects as tools for self-harm. The contrasting mind of a younger Elaine experiencing bullying is profiled by the grammatical normalization of self-harm and metaphorical mappings that profile idiosyncratic conceptual domains desired by the terrorized conceptualizer. I suggest that the linguistically foregrounded features in the paintings' descriptions allow the reader to establish an intratextual interrelation to the memories that have triggered the art. Art then becomes a reference point for the reader to juxtapose the protagonist's minds that consistently play off each other in ekphrastic writing and foreground cognitive limitations developed as a result of trauma.

References

Armstrong, C., U. Langås (2020), 'Introduction: Encounters between trauma and ekphrasis, words and images', in C. Armstrong, U. Langås (eds), *Terrorizing Images: Trauma and Ekphrasis in Contemporary Literature*, 1–12. Berlin: Walter de Gryuter.

Atwood, M. (1986), 'That certain thing called the girlfriend', *New York Times* [online]. Available online: https://archive.nytimes.com/www.nytimes.com/books/00/09/03/specials/atwood-girlfriend.html (accessed 15 November 2022).

Atwood, M. ([1988] 1990), *Cat's Eye*, London: Virago Press.
Banerjee, C. (1990), 'Atwood's time: hiding art in *Cat's Eye*', *Modern Fiction Studies*, 36 (4): 513–23.
BBC Bookclub (1999), *Margaret Atwood* [Sound recording]. Available online: https://www.bbc.co.uk/sounds/play/p00fpv3c (accessed 15 November 2022).
Brooks Bouson, J. (1993), *Brutal Choreographies: Oppositional Strategies and Narrative Design in the Novels of Margaret Atwood*, Amherst: The University of Massachusetts Press.
Cooke, N. (1992), 'Reading reflections: the autobiographical illusion in *Cat's Eye*', in M. Kadar (ed.), *Essays on Life Writing: From Genre to Critical Practice*, 162–70. Toronto: University of Toronto Press.
Diedrich, L. (2018), 'PTSD: A new trauma paradigm', in R. Kurtz (ed.), *Trauma and Literature*, 83–94. Cambridge: Cambridge University Press.
Dvořák, M. (2001), 'Margaret Atwood's *Cat's Eye*: Or the trembling canvas', *Études anglaises*, 54 (3): 299–309.
Emmott, C. and M. Alexander (2015), 'Defamiliarization and foregrounding: Representing experiences of change of state and perception in neurological illness autobiographies', in V. Sotirova (ed.), *The Bloomsbury Companion to Stylistics*, 289–307. London: Bloomsbury.
Fisher, L. (2015), *Between Oneself and the Other: Empathy, Dialogism, and Feminist Narratology in Two Novels by Margaret Atwood*, MRes thesis: Macquarie University.
Fludernik, M. (1996), *Towards a 'Natural' Narratology*, London: Routledge.
Fowler, R. (1977), *Linguistics and the Novel*, London: Methuen.
Gavins, J. (2007), *Text World Theory*, Edinburgh: Edinburgh University Press.
Gavins, J. (2012), 'Leda and the stylisticians', *Language and Literature*, 21 (4): 345–62.
Giovanelli, M. (2013), *Text World Theory and Keats' Poetry: The Cognitive Poetics of Desire, Dreams and Nightmares*, London: Bloomsbury Academic.
Giovanelli, M. (2018), '"Something happened, something bad": Blackouts, uncertainties and event construal in *The Girl on the Train*', *Language and Literature*, 27 (1): 38–51.
Giovanelli, M. (2022), *The Language of Siegfried Sassoon*, Cham: Palgrave Macmillan.
Giovanelli, M. and C. Harrison (2018), *Cognitive Grammar in Stylistics: A Practical Guide*, London: Bloomsbury.
Givner, J. (1992), 'Names, faces and signatures in Margaret Atwood's "Cat's Eye" and "The Handmaid's Tale"', *Canadian Literature*, 133: 56–75.
Harrison, C. (2017), 'Finding Elizabeth: Construing memory in *Elizabeth is missing* by Emma Healey', *Journal of Literary Semantics*, 46 (2): 131–51.
Herman, J. (1992), *Trauma and Recovery: The Aftermath of Violence from Domestic Abuse to Political Terror*, New York: Basic Books.
Hollander, J. (1988), 'The poetics of ekphrasis', *Word & Image*, 4 (1): 209–19.
Hubbard, K. (1989), 'Reflected in Margaret Atwood's *Cat's Eye*, girlhood looms as a time of cruelty and terror', *People*. 6 March, 205–6.

Kurtz, R. (2018), 'Introduction', in R. Kurtz (ed.), *Trauma and Literature*, 1–17. Cambridge: Cambridge University Press.

Lambrou, M. (2014), 'Narrative, text and time: Telling the same story twice in the oral narrative reporting of 7/7', *Language and Literature*, 23 (1): 32–48.

Langacker, R. W. (2008), *Cognitive Grammar: A Basic Introduction*, Oxford: Oxford University Press.

Leech, G. N. and M. Short ([1981] 2007), *Style in Fiction: A Linguistic Introduction to English Fictional Prose*, 2nd edn. New York: Pearson Longman.

Leys, R. (2000) *Trauma: A Genealogy*, Chicago, IL: University of Chicago Press.

Luckhurst, R. (2008), *The Trauma Question*, London: Routledge.

Mason, J. (2019), *Intertextuality in Practice*, Amsterdam: John Benjamins.

Nuttall, L. (2018), *Mind Style and Cognitive Grammar: Language and Worldview in Speculative Fiction*, London: Bloomsbury.

Osborne, C. (1994), 'Constructing the self through memory: "Cat's Eye" as a novel of female development', *Frontiers*, 14 (3): 95–112.

Paiva, A. B. (2022), 'Trauma and the fictional self-portrait in Margaret Atwood's *Cat's Eye* and Ana Teresa Pereira's *As Rosas Mortas*', *Anglo Saxonica*, 20 (1): 1–13.

Panagiotidou, M.-E. (2016), '"In the mind's eye": A cognitive linguistic re-construction of WD Snodgrass' "Matisse: The red studio."', *Language and Literature*, 25 (2): 130–44.

Panagiotidou, M.-E. (2017), 'Ekphrasis, cognition and iconicity: An analysis of W.D. Snodgrass's "Van Gogh: 'The starry night'"', in A. Zirker, M. Bauer, O. Fischer and C. Ljungberg (eds), *Dimensions of Iconicity*, 135–50. Amsterdam: John Benjamins.

Panagiotidou, M.-E. (2022a), 'Paradise lost: Cognitive grammar, nature, and the self in Diane Seuss's ekphrastic poetry', *Journal of World Languages*: 1–22.

Panagiotidou, M.-E. (2022b), *The Poetics of Ekphrasis*, Cham: Palgrave Macmillan.

Rimmon-Kenan, S. (2002), *Narrative Fiction: Contemporary Poetics*, 2nd edn. London: Routledge.

Rosenblatt, L. ([1938] 1970), *Literature as Exploration*, London: Heinemann.

Simpson, P. (1993), *Language, Ideology and Point of View*, London: Routledge.

Stockwell, P. (2009a), 'The cognitive poetics of literary resonance', *Language and Cognition*, 1 (1): 25–44.

Stockwell, P. (2009b), *Texture: A Cognitive Aesthetics of Reading*, Edinburgh: Edinburgh University Press.

Tabakowska, E. (2014), 'Point of view in translation: Lewis Carroll's Alice in grammatical wonderlands', in C. Harrison, L. Nuttall, P. Stockwell and W. Yuan (eds), *Cognitive Grammar in Literature*, 101–16. Amsterdam: John Benjamins.

Terkel, S. (1989), Interview with Margaret Atwood. *Studs Terkel Radio Archive* [Sound recording]. Available from: https://studsterkel.wfmt.com/programs/interview-margaret-atwood?t=513.74%2C573.068&a=YouKnowWe%2CIveGotLett (accessed 15 November 2022).

Verdonk, P. (2005), 'Painting, poetry, parallelism: Ekphrasis, stylistics and cognitive poetics', *Language and Literature* 14 (3): 231–44.
Vickroy, L. (2005), 'Seeking symbolic immortality: Visualizing trauma in *Cat's Eye*', *Mosaic* 38 (2): 129–43.
Webb, R. (2009), *Ekphrasis, Imagination and Persuasion in Ancient Rhetorical Theory and Practice*, Farnham: Ashgate.
Werth, P. (1999), *Text Worlds: Representing Conceptual Space in Discourse*, London: Longman.

10

Body, Mind, and Nature in Rossetti's 'For a Venetian Pastoral by Giorgione (In the Louvre)'

Maria-Eirini Panagiotidou

Introduction

Our contemporary understanding of the term 'ekphrasis' is closely tied to the definitions offered by Krieger and Heffernan during the last decade of the twentieth century as 'the imitation in literature of a work of plastic art' (Krieger 2019: 265) and 'the verbal representation of a visual representation' (Heffernan 1993: 3). Both definitions reflect an emphasis on the visual and may be considered within the larger context of 'the pictorial turn' (Mitchell 1994: 11) – namely the prevalence of images and pictures that has dominated our culture since the second half of the twentieth century. This emphasis on the visual was used by scholars to highlight a perceived competition between the word and the image termed 'the paragone' by Mitchell (1986). This antagonism and struggle for dominance between the two mediums is reflected in scholarly approaches to ekphrasis. For example, Heffernan notes that 'ekphrasis, then, is a literary mode that turns on the antagonism … between verbal and visual representation … To represent a painting or sculpted figure in words is to evoke its power … even as language strives to keep that power under control' (1993: 7).

Questions of power and control within the context of the paragonal relationship preoccupied much of the critical discussions in the last twenty years; however, more recently, theorists have started to embrace different theoretical foci that emphasize either the concept of 'encounter' (Kennedy 2012) or 'literary response' (Brosch 2018), redirecting the focus of ekphrasis on the collaborative, interactive relationship between its various agents (e.g. artwork, literary text, author, artist and reader). Brosch highlights this new approach when she states that 'ekphrasis emerges from a mode of articulation and its interaction with an audience – hence, the definition's emphasis on performance and response' (Brosch 2018: 227). The notions of 'interaction' and 'response' are critical to a new understanding of ekphrasis where readers emerge as equally important participants in the meaning-making process.

Interestingly, this shift towards reception is in line with the original conceptualization of ekphrasis by ancient rhetoricians as 'a descriptive (periēgēmatikos) speech which brings (literally "leads") the thing shown vividly (enargōs) before the eyes' (Webb 2009: 51). The descriptive quality of ekphrasis is considered together with the quality of *enargeia* (vividness), meaning that effective ekphrastic descriptions are characterized by the ability to vividly represent the subject in the listener's mind and to mentally transport them to the imagined scene. Enargeia was one of the primary means of persuasion whereby the audience could be captivated by the semblance of life language affords, feeling present at the events described and ultimately becoming emotionally engaged. These feelings of immersion and engagement were paramount to ancient writers and their understanding of ekphrasis.

For example, Zanker (1981) quotes Dionysus of Halicarnassus, an ancient Greek historian and teacher of rhetoric, who noted that enargeia transforms listeners into eyewitnesses; language offers them the ability to see the described event and feel the presence of the characters. In addition, Webb (2009) points out that Quintilian, the famous Roman rhetor, discusses enargeia in conjunction with an effect called *metastasis* or *metathesis* ('a transference [of time]') that allows the audience to be transported to either the past or the future. *Metathesis* goes beyond a simple switch to future or past events; it is critically connected to the orator's ability to make their audience feel as if they were present at the events described.

Ekphrasis and its effects on the reading experience, namely the sense of vividness (enargeia) and transference (metathesis), bring to mind similar considerations regarding literary effects in readerly perception within the broad field of literary studies. More specifically, the readerly feeling of immersion and the textual and psychological parameters that facilitate an immersive reading experience have attracted substantial critical attention in the last decades. Although different terminology has been used to describe the phenomenon, including the terms 'transportation' (Gerrig 1993), 'fictional recentering' (Ryan 2001), 'presence' (Kuzmičová 2012) and 'immersion' (Stockwell 2009), the overall emphasis and interest of literary scholars, stylisticians and psychologists remain constant: how readers become engaged with literature. In recent years, empirical studies (Nuttall and Harrison 2020; Kuijpers 2022) have investigated the way unprompted readers express their engagement with the text and have uncovered that they do rely heavily on embodiment and embodied metaphors to discuss their experiences.

In Nuttall and Harrison's study, which examined a sample of 200 reader reviews of the novel *Twilight* collected from Goodreads, the conceptual metaphor READING IS A JOURNEY emerges as the most widely used in their data set, followed by READING IS CONTROL, READING IS EATING and READING IS INVESTEMENT. Using a single review of the book *Bet me* by Jennifer Crusie from Goodreads, Kuijpers presents a case study where she traces the reader/reviewer's immersion into the story using text world theory. Although the conceptual metaphor READING IS TRANSPORTATION does not explicitly appear in the analysis, Kuijpers demonstrates how readers 'can show their movement from one text-world to another during reading' (Kuijpers 2022: 126) through the use of specific language choices. The social context of online reader reviews and their social discourse allowed the reviewer to express their level of absorption and convey what the reading experience was like. Kuijpers notes that, in an earlier study with David Miall (Kuijpers and Miall 2011), they found that readers who scored higher on the absorption scale were also more likely to report bodily feelings during reading. According to Kuijpers and Miall, this was due to the fact that readers 'exchange their own feelings and bodily awareness for those of the characters they read about, implying a deictic alignment or projection on the part of the reader' (Kuijpers 2022: 126).

This comment raises related questions to Kuzmičová's (2012) and Caracciolo's (2014; 2018) research on enactivism and embodied cognition. As mentioned above, Kuzmičová used the term 'presence' to describe the reader's immersion in a literary text which she defines as 'the sense of having physically entered a tangible environment, of "being there"' (2012: 24). She argues in favour of an embodied engagement with the literary work which arises from 'a first-person, *enactive* process of sensorimotor simulation/resonance' (2012: 24, emphasis in original). The reader's embodied presence in the world of the literary text is not just a matter of imagination, but it actively involves their bodily perceptions. In particular, textual descriptions of bodily movement will draw the reader into the text and trigger a 'flash of sensorimotor unity' (Kuzmičová 2012: 29) with the character. Caracciolo has made similar claims when looking at degrees of embodiment in literary reading. He has suggested that 'readers enact the storyworld by relying on the virtuality of their movements' (Caracciolo 2014: 100).

Drawing on work by researchers like Rolf Zwaan (2004; 2014) on the embodiment of language comprehension, Caracciolo argues that readers may enact a bodily perceptual experience in a narrative by simulating bodily movement. He, nevertheless, has proposed that the reader's sense of embodiment is not an all-or-none but rather a graded phenomenon that may range from limited embodied involvement to a fully fledged immersive experience. Weak embodied simulation stands in the middle of the scale; in this case, the readers' sense of embodiment does not have lasting effects on the reading experience and is forgotten soon afterwards.

In this chapter, I follow Caracciolo's claim regarding the presence of degrees of embodiment and explore what I call the 'transportational potential' of a literary text. Caracciolo's discussion has focused on bodily imagery and the readers' engagement with the perceptual experiences and physical actions of the characters. I present an expanded list of textual features that, when attended to by the reader, contribute to the feeling of immersion and presence in the world of the text. Returning to the concept of enargeia in the context of an ekphrastic poem by Dante Gabriel Rossetti titled 'For a Venetian Pastoral by Giorgione (In the Louvre)' (Rossetti 2003), I will explore the emergence of a fully embodied reading experience.

The pastoral tradition and Rossetti's sonnet

Rossetti's 'For a Venetian Pastoral by Giorgione (In the Louvre)' first appeared in the Pre-Raphaelite magazine *The Germ* in 1850 and was substantially revised before its publication in *Poems* in 1870. In this chapter, I analyse the revised version. The sonnet was inspired by a painting titled *Le Concert Champêtre* (Concert in the Open Air) that Rossetti saw in the Louvre in 1849. In a letter to his brother, Rossetti admits to being so taken by the work that he felt the need to sit down before it and compose the poem. The painting was originally attributed to Venetian painter Giorgione, as evidenced in Rossetti's title, but more recently, it has been considered the work of another Venetian master, Titian, who was Giorgione's contemporary (Marlow 2003). It depicts a pastoral scene set in an idyllic countryside setting. In the centre of the composition, two male musicians clothed in Renaissance-style attire engage in singing. They are framed by two nude female figures: the one on the left is holding a pitcher and is pouring water into a fountain while the one on the right has her back turned to the viewer and is holding a flute. In the background, one may discern the figure of a shepherd with his herd. Brown (2006) identifies Titian's painting as the epitome of the revived pastoral genre which placed a particular emphasis on the role of landscape, similarly to its literary counterpart. In 1495, some years before the painting's composition (*c.* 1510), the first complete works of the Hellenistic poet Theocritus titled *Idylls* were printed in Venice. Together with Virgil's *Eclogues*, they exerted a major influence on artistic production, leading to the publication of Jacopo Sannazaro's *Arcadia* in 1502, a collection of prose and poetry written in Italian. The fact that Sannazaro's work was published in the vernacular rather than in Latin facilitated its accessibility since both educated patrons and artists who did not have knowledge of Latin could read it (Brown 2006).

According to Conan (1997), a common thread unites the pastoral landscape as it was conceptualized by ancient Greek writers, Roman poets, Renaissance patrons and artists, all the way to contemporary urbanities: a love of nature connected with escapism from the thrills and dangers of urban life. Idealized descriptions of landscapes where humans, deities, nature and animals co-existed harmoniously abound in pastoral literature and paintings. This feeling of harmonious co-existence may also be observed in Titian's canvas, exemplified by the presence of the two male figures. Brown notes that the elaborate clothes of the one musician indicate his status as an urban dweller while his companion's simpler attire and bare feet denote more humble origins; he is perhaps a shepherd in a 'dual role of herdsman … and musician or poet' (Brown 2006: 19). The reconciliation of oppositions 'lies at the heart of the pastoral idiom', facilitating 'the dialectic between city and country, cultured and rustic, art and nature' (Brown 2006: 19).

Returning to the poem, Rossetti has chosen a variation of the Petrarchan sonnet but does not follow the four-part division of the strict form (Ireland 1979), nor does he allocate equal attention to the four main figures of the canvas. Beginning with a vague address to the female figure on the left, who pours water out of the pitcher she is holding, the description then moves to the distant landscape before returning to the hand of the male musician. Almost immediately, the focus moves to the second female figure with the poetic voice posing a set of questions regarding her behaviour that ultimately remain unanswered. In the final tercet, the poetic voice shifts away from pictorial description and directly addresses the reader/viewer, pleading with them not to disrupt the second woman's trance before concluding with expressing a wish for 'infinite prolongation of the depicted scene' (Ireland 1979: 308):

'For a Venetian Pastoral by Giorgione (In the Louvre)'

Water, for anguish of the solstice: – nay
But dip the vessel slowly, – nay, but lean

And hark how at its verge the wave sighs in
Reluctant. Hush! beyond all depth away
The heat lies silent at the brink of day:
Now the hand trails upon the viol-string
That sobs, and the brown faces cease to sing,
Sad with the whole of pleasure. Whither stray
Her eyes now, from whose mouth the slim pipes creep
And leave it pouting, while the shadowed grass
Is cool against her naked side? Let be: –
Say nothing now unto her lest she weep,
Nor name this ever. Be it as it was, –
Life touching lips with Immortality.

Previous critical discussions of the poem have focused on its connection with the broader ekphrastic tradition. Recently, Helsinger (2022) focused on the presence and role of music in six ekphrastic sonnets by Rossetti, including the one discussed in this study. She suggests that music becomes 'a special kind of embodied knowing' (Helsinger 2022: 218) with the poem evoking the experience of music through pictorial means. Listening – either to actual music or to unheard music, as in this poem – involves the senses, namely sight, hearing and touch, and Rossetti's sonnets become doubly ekphrastic by functioning 'as at once narrative evocations of an experience of music as feeling (sensory and emotional) and … as a figure for the tonal and rhythmic architecture of music as, like poetry, it is experienced in time' (Helsinger 2022: 218). The latter is achieved through linguistic description and repetition of visual patterns that in turn suggest harmonic and rhythmic patterns characteristic of musical compositions. In 'On a Venetian Pastoral by Giorgione', Rossetti wishes to capture the elusive moment where music has just ceased and to comment on its temporal elusiveness. The poem links together painting and music, and '[p]ainting's difficulties are for Rossetti an occasion to think about the temporal, sounded qualities of poetry, even – or especially – in an age of print' (Helsinger 2022: 226). Ireland makes a similar point, suggesting that Rossetti has recorded the moment 'between sound and silence' before 'the spell of the music has been broken' (Ireland 1979: 308).

In another recent paper, McGhee approaches the poem from the Renaissance art-theoretical concept of *vaghezza* ('vagueness') which brings together 'notions of haziness, beauty, mutability, and indefinability … l[ying] at the heart of [Rossetti's] poetic technique and artistic ambitions' (McGhee 2021: 281). According to McGhee, the vagueness and ambivalence that characterize the poetic voice's addresses and grammatical choices draw readers into the world of the poem/painting, only to hold them back again. Readers are encouraged to experience a sense of vivid temporality as the poetic voice employs language to capture 'own process of response to Giorgione's picture rather than in the realization of an iconic verbal equivalent' (McGhee 2021: 293).

Both McGee's and Helsinger's analyses touch upon the poem's ability to communicate a feeling of embodied experience in terms of a sense of temporality and music as embodied knowing, respectively. In my analysis of Rossetti's poem, I will explore the concept of embodied presence and demonstrate how the presence of deictic expressions, image schemas, conceptual metaphors, personification and lexical choices emphasizing bodily movement facilitate the reader's own sensorimotor involvement.

Enactment and bodily engagement

In the introductory section, I referred to enactivism, a term originally proposed by Varela, Thompson and Rosch in their influential work *The Embodied Mind* (1991). Rejecting the strict objectivism of first-generation cognitive science, they proposed that cognition arises from the interaction between an organism and its environment. In their application of the concept to literary analysis, Kuzmičová (2012) and Caracciolo (2014; 2018) explored how textual references to bodily movement may facilitate the reader's immersion in the world of the text. The reader's embodied presence in the world of the literary text is not just a matter of imagination: it actively involves their bodily perceptions. In his analysis of José Saramago's *Blindness*, Caracciolo (2014) argues that the involvement of the reader's own body through embodied simulations of described bodily actions allows them to engage with bodily perceptual experiences of characters. Importantly, readers 'can enact narrative space by simulating a hypothetical perceptual experience, even in the absence of fictional characters to which they could attribute that experience' (Caracciolo 2014: 103). Kuzmičová suggests that the sense of presence and immersion and the feeling that a reader has 'physically entered a tangible environment' may be 'achieved when certain forms of human bodily movement are rendered in the narrative, as compared to when they are not' (2012: 25). She goes on to propose that transitive movements that involve everyday objects, like books, beer cans, or hammers, have a higher immersive potential, noting that more familiar objects afford a stronger sense of immersion to the reader.

Turning our attention to the poem, bodily movements are particularly emphasized in the initial lines. These lines are also characterized by a high degree of ambiguity; as McGhee (2021) notes, it is not immediately clear by whom they are articulated and to whom they are addressed. The poetic voice could be coming from inside or outside the painting, being one of the depicted figures, another observer occupying the same space, or a viewer outside the frame. At the same time, the lines may address the figure pouring water from the pitcher, or they could also be read as 'as dramatic speech spoken by a figure in the painting' (McGhee 2021: 288). While these multiple interpretive possibilities may initially cause some processing difficulty, it can also be suggested that the ambiguous address works together with the verbs of motion to engage the reader with the poem. In line 2, the presence of the imperatives 'dip the vessel slowly' and 'lean' engage the reader's body and invite them to the simulate their own experience of performing these actions. The verb 'dip' describes a transitive bodily movement that is directed to a specific object in the proximity, namely 'the vessel'.

As discussed in the previous paragraph, according to Kuzmičová, the use of the transitive verb phrase may elicit enactive responses and an increased sense of presence, prompting the reader to enact the process of dipping the pitcher into water since movements affecting everyday artefacts a strong 'immersive potential in terms of resonance and multimodal imagery' (2012: 31). In addition to action verbs that indicate strong bodily involvement, the poem also contains verbs that invite the reader to simulate other bodily processes, specifically connected to the senses and the act of speaking. In line 3, the reader/addressee is instructed to 'hark' and listen to the sound of the water touching the mouth of the pitcher, while in line 4, the imperative 'hush' alludes to the prohibition of bodily movement. The addressee is asked to refrain from speaking, a plea repeated thrice more in subsequent lines: in line 11, they are asked not to disturb the scene and the figures of the painting but to 'let be', a request that is further elaborated in line 12 when the poetic voice asks them to 'say nothing now' to the second female figure so as not disrupt her reverie. Then in the next line, the prohibition is expanded to any future references to this encounter: 'Nor name this ever' (l. 13).

Although the use of three consecutive imperative forms together with 'hush' in line 4 semantically indicates the lack of movement of our articulators and inaction, the process of *negative accommodation* (e.g. Giovanelli 2013) may activate the bodily act of speaking. For example, the negative constructions 'say nothing' and 'nor name this' trigger a scenario where the speaker does perform these actions. As a result, the act of speaking is evoked only to be immediately negated.

Despite the short form of the sonnet, the lines contain multiple references to bodily movement and actions, which may in turn encourage the reader's own bodily engagement and sense of presence through the process of mental simulation. The high occurrence of these references in the initial and final lines may act as a means of reinforcing the reader's immersion in and engagement with the poem.

Deixis and the feeling of presence

In addition to expressions that indicate bodily involvement, deixis may also facilitate the reader's transportation to the world of the poem. *Deixis* (from the Greek word δείκνυμι 'deiknymi' meaning 'to show') refers to a set of linguistic expressions that encode an utterance within a particular spatiotemporal, perceptual, discoursal and social context. Stylisticians (e.g. Gibbons and Whiteley 2018; Stockwell 2020) have connected deixis to our embodied position in the world, noting that our understanding of deictic expressions like 'I', 'this', 'now', 'here' encodes a speaker's particular perspective and should be interpreted based on context-sensitive parameters. Specifically, these expressions point to the utterance's zero-point, or *origo*, (Bühler 2011) which may be defined as the deictic centre determined by the speaker's embodied subjectivity, capturing their viewpoint and perception of the world. When it comes to reading literary texts, readers are capable of projecting their deictic centre into the storyworld, thus allowing them to adopt viewpoints of characters and become immersed in the narrative. This is known as *deictic projection* and was proposed within the context of deictic shift theory.

In Rossetti's poem, several types of deictic expressions may prompt the reader to perform a *deictic shift* (Stockwell 2020) and enter the fictional world of the poem. In terms of perceptual deixis which captures 'the perceptive participants in the text, encoded in pronouns, demonstratives, definite articles and definite reference' (Stockwell 2020: 54), multiple elements facilitate the shift, including the inferred addressee 'you', definite noun phrases and the demonstrative 'this' in line 13. The repeated use of the imperative form in the initial lines may urge the reader to assume the role of the protagonist/addressee and mentally simulate these actions, as we saw in the previous section. Although the imperative in English lacks a visible subject, the pronoun 'you' is typically inferred and is assumed to occupy the subject position. In the context of this poem, the 'you referent' is at the same time addressed extradiegetically but is also invited to act as the story's protagonist; this direct address may have a deictic function since it assumes the presence of the addressee in the communicative situation (see also Green 1992). Also, the poem is rich in definite descriptions of objects and figures, suggesting they are immediately available to the speaker and addressee and close to their environment. For instance, pictorial elements are introduced using the definite article: 'the vessel' (l. 2), 'the hand', 'the viol-string' (l. 6) 'the brown faces' (l. 7), 'the slim pipes' (l. 9), 'the shadowed grass' (l. 10). The consistent use of the definite article creates the impression that the elements are readily available for processing and helps maintain a sense of immersion.

In addition to perceptual deixis, temporal deixis, which identifies the temporal context of a text, is foregrounded in the central part of the sonnet through the use of the temporal adverb 'now' in lines 6, 9 and 12. The repetition of the adverb anchors the discourse in the present moment and encourages the reader's immersion into the world of the poem. The events are portrayed as unfolding in the present moment, and the poetic voice continuously brings this to the attention of the reader/viewer. In conjunction with the imperatives and the exclamation 'nay' (also repeated twice), the sense of immediacy as well as urgency is amplified, and the reader is invited to assume the role of the addressee and be absorbed in the narrative.

Conceptual metaphors, image schemas, embodiment and nature

The embodied view of cognition expressed in Varela, Thompson and Rosch's work (1991) is also the fundamental premise of conceptual metaphor theory (CMT) initially proposed by Lakoff and Johnson in their seminal work *Metaphors We Live By* (1980). CMT has since been elaborated and developed by numerous scholars, including Johnson (1987), Lakoff and Turner (1989) and Gibbs (1994). According to CMT, our conceptual system is grounded in our perceptions and interactions with the environment and metaphorical language reflects our bodily experiences and psychological processes. *Conceptual metaphors* arise from *crossdomain mappings* between a concrete *source domain*, which forms part of our everyday experiences, and an abstract *target domain*, which is conceived in terms of a more physically and perceptually concrete concept, i.e., the source domain. For instance, the primary conceptual metaphor AFFECTION IS WARMTH, which underlies linguistic expressions like 'He is a warm person' or 'They are warming up to the new professor', is established based on crossdomain mappings between the concrete, experiential domain of warmth and the more abstract domain of affection. As Lakoff and Johnson (1980; 1999) have pointed out, this CM may be motivated by the correlation between affection and warmth experienced by young children when they are held affectionately by their parents.

Conceptual metaphors play a significant role in Rossetti's poem, not only reinforcing a sense of presence but also creating a link between the natural world and its human inhabitants. A close reading reveals the presence of numerous personifications throughout the poem whereby natural elements, objects and abstract concepts are attributed human qualities. In line 3, the wave 'sighs' while in line 5 the heat 'lies silent'; later on, the viol-string 'sobs' (l. 7) and the pipes 'creep' (l. 9), and in the last line 'Life is touching lip with Immortality'. The first two examples express the CM NATURE IS A PERSON, using verbs and adjectives which typically take humans as their subject ('sighs', 'lies') or evaluate human behaviour ('silent') to metaphorically describe the landscape and the weather. In particular, 'the heat lies silent' may be connected to the CM WEATHER IS A HUMAN BODY (Goatly 2007), providing agency to and humanizing the elements. Goatly mentions that '[t]raditionally landscape and weather have been viewed as inanimate and incapable of agency, landscape especially' (2007: 123), but in Rossetti's poem, humans and the landscape are seen as approaching each other, thus emphasizing their harmonious co-existence, a predominant characteristic of the pastoral as we saw previously.

Interestingly, as the poem progresses, personification is extended to human-made objects and abstract notions: the viol is described as sobbing, a human emotional response, and the pipes as creeping, another action typically performed by humans. These descriptions make use of the CM INANIMATE OBJECTS ARE PEOPLE and function as a manifestation of the interconnected relationship between the different elements of the canvas. Inanimate objects are not depicted as detached from human influence but exhibit human traits, reinforcing the correspondences between the two. Personification is also employed in the last line where it reaches its climactic point. The abstract notions 'life' and 'immortality' are capitalized and presented as having human body parts ('lips') and engaging in a human activity ('touching lips'). The line also blurs the boundaries between the physical world and the world of abstract ideas and serves as a final iteration of the poem's powerful tendency to show an affinity between human and non-human entities.

The CM IDEAS ARE PEOPLE underlies this representation and further facilitates the reader's engagement with the entities inhabiting the poem through a process comparable to Stockwell's notion of empathetic recognizability (2009: 25). Empathetic recognizability postulates that textual elements are more likely to attract the reader's attention if they denote human agency, whereas abstractions are less likely to be attended to. The prevalent use of personification in the poem may have a similar effect as it encourages readers to engage with abstractions and objects as if they were human entities. If this identification is possible, the effect may have broader implications in terms of enhancing the reader's sense of presence. More specifically, it may involve an enactive response arising from the multiple references to bodily movements, reinforcing the reader's experiential involvement.

In addition to personification, the interconnectedness between humans and their broader environment is highlighted through the presence of an underlying image schema. Image schemas are preconceptual structures that arise from our everyday perceptual and motor experiences, acting as '"distillers" of spatial and temporal experiences' (Oakley 2007: 215). The term first appeared in the works of Lakoff (1987) and Johnson (1987), and some examples include but are not limited to the following: PATH/SOURCE-PATH-GOAL, CENTER-PERIPHERY, BALANCE, RESISTANCE, CONTAINER/CONTAINMENT, UP-DOWN, CONTACT and SCALE. Image schemas may also provide underlying structures for CMs as in the example of RELATIONSHIPS ARE JOURNEYS, whose source domain is primarily structured by the SOURCE-PATH-GOAL schema.

In Rossetti's poem, one may observe a prominent presence of the CONTACT image schema, one of the primary schemas identified by Johnson (1987) that show a clear link with physical experience. The schema underlies the use of prepositions like *on* or *against* to convey contact in sentences like 'The book is on the table' and 'The suitcase is against the wall'. As the poem unfolds, the close connection between human and non-human entities is alluded to and the interaction and communion among the different elements of the canvas are conveyed at various points. In line 3, water makes contact with the edge of the jar, and later on the hand of one musician is depicted as 'trailing upon the violstring' while the pipe is shown is the process of being separated from the nymph's lips. Additionally, in lines 10–11, the preposition 'against' profiles the contact between the naked skin of the nymph and the natural environment that surrounds her, and in the final line, contact is once again alluded to through the phrase 'touching lips'. In conjunction with personification, the CONTACT image schema may act as a reminder of the embodied presence of not only the protagonists but also the reader; bodily experiences are linked to perceptual observations of contact, thus strengthening the experiential quality of immersion.

Conclusion

This chapter explored the concept of a text's transportational potential which hinges upon the identification of a series of linguistic and cognitive structures. The presence of these features in a text may facilitate the reader's immersion into the fictional world. As we saw in the introduction to the chapter, the concept of readerly immersion has been explored from various perspectives; the current study proposes that immersion should be seen as a graded instead of all-or-none phenomenon whose emergence depends on the presence of the textual features outlined above as well as the reader's ability to attend to them. At one end of this continuum, if a text is rich in descriptions of bodily movements and actions that elicit enactive responses, deictic expressions that facilitate deictic projection, and image schemas and CMs that invite the reader to simulate a bodily action or state, it is characterized by high transportational potential. A significant occurrence of these features may prompt the reader to imaginatively experience their own presence in the fictional world and to sustain this sense of immersion.

In Rossetti's poem, vivid and recurrent references to bodily movements and high-embodiment descriptions, together with the repeated use of deictic expressions function as textual cues that facilitate the reader's projection into the world of the painting. Multiple references to bodily actions such as 'lean', 'dip', 'hark' and 'hush' engage the reader's body and allow them to experience these motions. Kuzmičová (2012) has argued that a stable sense of presence is elicited if references to bodily movement are evenly distributed throughout the text. Therefore, texts with a high degree of transportational potential exhibit this feature, as can be observed in Rossetti's poem where the bodies of the addressee and of the painting's subjects are continuously addressed. Additionally, the underlying CONTACT image schema further supports the reader's bodily engagement, acting as a reminder of the connection between the body and the environment. Haptic perception and the active exploration of objects and surroundings are actively invoked in the poem. The importance of touch and contact is highlighted throughout the text implicitly and explicitly, culminating in the final line where the abstractions are personified and united in harmony.

While some texts have a high transportational potential, others may be situated in the middle of the continuum, where the immersive effect is seen as more transitory. The transitory nature of readerly immersion may result from some of the identified features being absent or present only to a partial degree, as well as being unevenly distributed throughout the text. At the other end of the continuum, we encounter texts with limited transportational potential. In this case, the textual cues are largely absent, and their sparse occurrence is unlikely to be attended to by the reader and facilitate their transportation to the world of the text. It should be noted here that the presence or absence of these cues is only one aspect influencing immersive reading experiences. Equally important are the readers' own predispositions, personal memories and emotional responses. In particular, the reader's emotional reactions to a text and ability to empathize with fictional characters have been found to be causative of the presence of transportation (Dixon and Bortolussi 2017).

Rossetti's sonnet demonstrates how ekphrasis and enargeia are intrinsically connected and how feelings of presence and engagement may arise with the poem moving beyond description to achieve this. Ireland has pointed out that the 1850 version is 'more faithful to the painting in its interpretation' (Ireland 1979: 311), containing more references to colours and pictorial elements while the 1870 revision seems to move towards abstraction. Nevertheless, I want to suggest that the 1870 sonnet analysed here is primarily preoccupied with immersing the addressee/reader into the poetic universe and helping them perceive the painter's intended purpose thematically and artistically. In other words, the revised version focuses on communicating a perfect moment of harmony, resonating *Le Concert Champêtre*'s overarching theme. It is no coincidence that when discussing the artwork, Marlow notes that the painting itself may be interpreted as 'a poetic union of Man and Nature ... suggested by the integration of the human forms with the shapes of the landscape' (2003: n.p.). The poet wishes to communicate this union and create a poetic equivalent of the painting. The integration pervades the grammar of the poem and is communicated through the widespread presence of the CONTACT image schema, together with the prevalence of ontological metaphors involving personification of objects, natural elements, and abstractions that draw attention to the reader's/addressee's physical body and their sensory-motor presence. The CMs NATURE IS A PERSON, INANIMATE OBJECTS ARE PEOPLE and IDEAS ARE PEOPLE allow the reader to engage with non-human entities in equal terms and consider their harmonious co-existence in a fictional environment where the abstract and concrete, ideal and real, art and life co-occur. The reader is invited to contemplate the bond between humans and their environment and how all elements are intrinsically connected. Though a close description of the painting is absent, the idyllic atmosphere of the pastoral is echoed in the lexis and grammar, creating a transcendental experience for the reader.

References

Brosch, R. (2018), 'Ekphrasis in the digital age', *Poetics Today*, 39 (2): 225–43.
Brown, D. A. (2006), 'Venetian painting and the invention of art', in D. A. Brown, S. Ferino Pagden, J. Anderson and B. Hepburn Berrie (eds), *Bellini, Giorgione, Titian, and the Renaissance of Venetian Painting*, 15–37. ed. by Washington/Vienna: National Gallery of Art; Kunsthistorisches Museum, in association with Yale University Press. New Haven and London.
Bühler, K. (2011), *Theory of Language: The Representational Function of Language*, trans. Donald Fraser Goodwin and Achim Eschbach, Amsterdam/Philadelphia: John Benjamins Pub. Co.
Caracciolo, M. (2014), *The Experientiality of Narrative: An Enactivist Approach*, Berlin/Boston: De Gruyter.
Caracciolo, M. (2018), 'Degrees of embodiment in literary reading: Notes for a theoretical model, With American psycho as a case study', in S. Csábi (ed.), *Expressive Minds and Artistic Creations: Studies in Cognitive Poetics*, 11–31. New York: Oxford University Press.
Conan, M. (1997), 'Poetry into landscape evolving views of the pastoral in painting and poetry from antiquity to the nineteenth century', *The Journal of Garden History*, 17 (3): 165–70.
Dixon, P. and M. Bortolussi (2017), 'Elaboration, emotion, and transportation: Implications for conceptual analysis and textual features', in F. Hakemulder, M. M. Kuijpers, E. S. Tan, K. Bálint and M. M. Doicaru (eds), *Narrative Absorption*, 199–215. Amsterdam: John Benjamins Pub. Co.
Gerrig, R. J. (1993), *Experiencing Narrative Worlds: On the Psychological Activities of Reading*, New Haven: Yale University Press.
Gibbons, A. and S. Whiteley (2018), *Contemporary Stylistics: Language, Cognition, Interpretation*, Edinburgh Textbooks on the English Language – Advanced, Edinburgh: Edinburgh University Press.
Gibbs, R. W. (1994), *The Poetics of Mind: Figurative Thought, Language, and Understanding*, Cambridge [England]/New York: Cambridge University Press.
Giovanelli, M. (2013), *Text World Theory and Keats' Poetry: The Cognitive Poetics of Desire, Dreams and Nightmares*, London/New York: Bloomsbury Academic.
Goatly, A. (2007), *Washing the Brain: Metaphor and Hidden Ideology* [Discourse Approaches to Politics, Society, and Culture], Amsterdam/Philadelphia: John Benjamins Pub. Co.
Green, K. (1992), 'A study of Deixis in relation to lyric poetry' (unpublished Doctoral Dissertation, University of Sheffield).
Heffernan, J. A. W. (1993), *Museum of Words: The Poetics of Ekphrasis from Homer to Ashbery*, Chicago: University of Chicago Press.
Helsinger, E. (2022), 'Picturing music: Doubling ekphrasis in six Rossetti sonnets', *Journal of Victorian Culture*, 27 (2) (2022): 216–35.

Ireland, K. R. (1979), 'A kind of pastoral: Rossetti's versions of Giorgione', *Victorian Poetry*, 17 (4): 303–15.

Johnson, M. (1987), *The Body in the Mind: The Bodily Basis of Meaning, Imagination, and Reason*, Chicago: University of Chicago Press.

Kennedy, D. (2012), *The Ekphrastic Encounter in Contemporary British Poetry and Elsewhere*, Farnham: Ashgate.

Krieger, M. (2019), *Ekphrasis: The Illusion of the Natural Sign*, 2nd edn. Baltimore: Johns Hopkins University Press.

Kuijpers, M. M. (2022), 'Bodily involvement in readers' online book reviews: Applying text world theory to examine absorption in unprompted reader response', *Journal of Literary Semantics*, 51 (2) (2022): 111–29.

Kuijpers, M. M. and D. Miall (2011), 'Bodily involvement in literary reading: An experimental study of readers' bodily experiences during reading', in F. Hakemulder (ed.), *De Stralende Lezer: Wetenschappelijk Onderzoek Naar de Invloed van Het Lezen*, 160–82. Delft: Eburon.

Kuzmičová, A. (2012), 'Presence in the reading of literary narrative: A case for motor enactment', *Semiotica*, 189/1/4, 23–48.

Lakoff, G. (1987), *Women, Fire, and Dangerous Things: What Categories Reveal about the Mind*, Chicago: University of Chicago Press.

Lakoff, G. and M. Johnson (1980), *Metaphors We Live By*, Chicago: University of Chicago Press.

Lakoff, G. and M. Johnson (1999), *Philosophy in the Flesh: The Embodied Mind and Its Challenge to Western Thought*, New York: Basic Books.

Lakoff, G. and M. Turner (1989), *More than Cool Reason: A Field Guide to Poetic Metaphor*, Chicago: University of Chicago Press.

Marlow, K. (2003), 'Fête champêtre', *Grove Art Online*. Available online: http://www.oxfordartonline.com/groveart/documentID/oao-9781884446054-e-7000028098 (accessed 26 July 2022).

McGhee, F. (2021), 'Rossetti's Giorgione and the Victorian "Cult of Vagueness"', *The Cambridge Quarterly*, 50 (3): 279–95.

Mitchell, W. J. T. (1986), *Iconology: Image, Text, Ideology*, Chicago: University of Chicago Press.

Mitchell, W. J. T. (1994), *Picture Theory: Essays on Verbal and Visual Representation*, Chicago: University of Chicago Press.

Nuttall, L. and Ch. Harrison (2020), 'Wolfing down the Twilight series: Metaphors for reading in online reviews', in H. Ringrow and S. Pihlaja (eds), *Contemporary Media Stylistics*, 35–60. London: Bloomsbury Academic.

Oakley, T. (2007), 'Image schemas', in D. Geeraerts and H. Cuyckens (eds), *The Oxford Handbook of Cognitive Linguistics*, 214–35. Oxford: Oxford University Press.

Rossetti, D. G. (2003), *Collected Poetry and Prose*, edited by J. McGann, New Haven: Yale University Press.

Ryan, M.-L. (2001), *Narrative as Virtual Reality: Immersion and Interactivity in Literature and Electronic Media*, 1st edn, Baltimore: Johns Hopkins University Press.

Stockwell, P. (2020), *Cognitive Poetics: A New Introduction*, 2nd edn, London/New York: Routledge.

Stockwell, P. (2009), *Texture: A Cognitive Aesthetics of Reading*, Edinburgh: Edinburgh University Press.

Varela, F.J., E. Thompson and E. Rosch (1991), *The Embodied Mind: Cognitive Science and Human Experience*, Cambridge, MA: MIT Press.

Webb, R. (2009), *Ekphrasis, Imagination and Persuasion in Ancient Rhetorical Theory and Practice*, Farnham/Burlington: Ashgate.

Zanker, G. (1981), 'Enargeia in the ancient criticism of poetry', *Rheinisches Museum Für Philologie*, 124/3/4, 297–311.

Zwaan, R. A. (2004), 'The immersed experiencer: Toward an embodied theory of language comprehension', in B. H. Ross (ed.), *Psychology of Learning and Motivation*, 35–62. San Diego/London: Elsevier Academic Press.

Zwaan, R. A. (2014), 'Embodiment and language comprehension: Reframing the discussion', *Trends in Cognitive Sciences*, 18 (5): 229–34.

Conclusion: Pathways to ECO

Malgorzata Drewniok and Marek Kuźniak

In his Introduction to the present volume, Peter Stockwell has drawn a clear map of the contemporary transition from ECO-centric traditional stylistics to ECO-centric analyses, including radical ecological stylistics. As we demonstrate below, the present volume covers the insights that can be characterized as a midway between a traditional approach that employs conventional linguistic tools to explore various levels of text organization, with the exploration itself focused on self-consciousness, and most recent proposals that integrate the methodological apparatus with the principles of the so-called 4E cognition. In this respect, the collection is rightly labelled 'applied cognitive ecostylistics', as it applies the findings from cognitive studies to address the issues of ECO seen as a certain pattern of thinking about broadly conceived ecology of human life.

The marriage between cognitive linguistics and ecology seems natural. The fundamental link is shared human experience that already forms a platform of extended cognition. All this shifts the perspective from the exclusive 'I' to the inclusive 'we' not only at the level of rhetoric but primarily at the level of cognition. It seems that the studies within the Extended Conceptual Metaphor (Kövecses 2020) are the best illustration of the stage where we are as stylisticians. By going beyond the limits of our individual experience cognitive studies (with a particular role of cognitive linguistics) aim at intersubjectivity, or convergence of evidence about our human mind across disciplines. The project is open-ended at its roots. The integration with the ECO progresses with the big challenge ahead, i.e. how to reconceptualize and re-cast the well-established rigour of study in stylistics to meet the growing demands that arise from deep ecological postulates which appear as 'game changers' at the plane of the description of human ontology. This question remains unanswered, though, as we see below (Bogusławska et al. 2022), attempts to move from cognitivism to ecologism are already reported. At the level of epistemology, the move is from the 'worldly', perceptible, commonsensical and local facts established by Newtonian physics to invisible, non-intuitive and non-local quantum physics, where the statements made in the humanities are still only tentative proposals than facts. All in all, the fascinating journey from EGO to ECO has barely begun with no clear point of destination at hand.

Elżbieta Hoły-Łuczaj, in her work on radical non-anthropocentrism (2018: 78), attempts to synthesize the main points raised by the philosophers working in the paradigm of deep ecology. As she observes (2018), in 1984, George Sessions and Arne Naess summarized their thoughts on deep ecology. They formulated eight principles of deep ecology as accessibly and unanswerably as possible, with the hope that they would meet with understanding and acceptance by people of different religious and philosophical orientations. These principles are as follows:

1. The prosperity and development of human and non-human life on Earth are values in themselves regardless of the usefulness of non-human life forms to humans.
2. The abundance and diversity of life forms contribute to the realization of these values and are values in themselves.
3. Humans have no right to limit this richness and diversity except in special life situations.
4. The development of non-human forms of life requires the suppression of human population growth.[1]

Marta Bogusławska, Alina Andreea Dragoescu Urlica and Lulzime Kamberi, the editors of *From Cognitivism to Ecologism in Language Studies* (2022), suggest a similar radical move from the classical Western ontology anchored in materialism to post-ontology immersed in processualism. Bogusławska (2022) claims that the split is actually crystalized and practically unbridgeable. As she asserts:

> In order to delineate the theoretical framework of the twenty-first-century ecolinguistics, we start from reflecting upon the general idea of the state of the facts we are about to observe and study. This is not an obvious discernment, as today we have two general choices, (i) we can decide on the classical world model/classical paradigm, or (ii) the postclassical world model/postclassical, post-Newtonian, holistic paradigm. Both theoretical, meta-cognitive choices are available to scholars and each carries very different consequences upon choosing one or the other.
>
> (Bogusławska 2022: 10)

In the same spirit, Govianni Aloi and Susan McHugh, the editors of *Posthumanism in Art and Science* (2021a), envision a large-scale transition into posthuman ontology and epistemology. In their Introduction to the volume, Aloi and McHugh (2021b) explore the effects of posthuman thinking and acting. On a philosophical plane, they pinpoint Jacques Derrida's (2002) famous article *The Animal That Therefore I am* and Michel Foucault's (1966) *The Order of Things*, the notion of *naturecultures* that informs about the actual blend of the two concepts seen so-far as dichotomous.[2] This provokes an anti-Anthropocene aesthetic in art and an anti-epistemological approach in philosophy, as best manifested in the works by Philippe Descola (e.g. 2014).

On a literary plane, Aloi and McHugh (2021b) synthesize the movement away from *Anthropos* by bringing to the reader the influential *Why Look at Animals?* by John Berger (2005), not to mention Cary Wolfe's 2010 work on *What Is Posthumanism*. Along these lines, Aloi and McHugh also note a transformational strand in posthuman philosophy by quoting Donna Haraway's 1985 *A Cyborg Manifesto* – another challenge to Eurowestern ontology of dualisms or Doo-Sung Yoo's 2010 *Organ-Machine Hybrid* – a far-reaching transformational manifesto that touches the roots of life processes.

Last but not least, in art and science, the effects of the posthuman seem evident. Aloi and McHugh (2021b) evoke, e.g. cubism as the flagship project of the anti-Baroque worldview. The authors also mention dOCUMENTA art exhibition in 2012, which was a good illustration of how posthumanism addresses the issues connected with the aspects of the environment and the related concerns. Certainly, the twentieth-century scientific heritage in physics as best embodied by Albert Einstein's theory of relativity and sub-atomic levels of awareness opened up by quantum physics, only speeded up the transition with the Anthropocene well on the decline.

Unlike the positions outlined above, our volume does not go as far as to propose such a substantial breakaway from anthropocentrism in favour of non-anthropocentric perspectives. In suggesting the need to emphasize the ego–eco continuum, we note that cognition is central to *Homo sapiens*. In essence, egocentricity in thinking and acting seems natural. However, as articles presented on the pages of the present volume have shown, the said egocentricity, if counterbalanced by the out perspective, loses much of its latent negative import embodied by the egoist stance. Such seen egocentricity leads us towards humanism hinged on the desired in–out synergy of life.

This synergy is achieved in ecolinguistic studies on many planes: social media communication and nature, media ecology, ecology of translation, ecological thinking, the language of landscape and the language of the animal world. Anthony Nanson (2021), in his deeply engaging *Storytelling and Ecology: Empathy, Enchantment and Emergence in the Use of Oral Narratives*, accentuates the need for the revival of oral storytelling as an important vehicle for reconnecting people in the contemporary era of individuation and alienation:

> Linking the ongoing ecological crisis with contemporary conditions of alienation and disenchantment in modern society, this book investigates the capacity of oral storytelling to reconnect people to the natural world and enchant and renew their experience of nature, place and their own existence in the world ... Detailed analysis of traditional, true-life and fictional stories shows how spoken narrative language can imbue landscapes, creatures and experiences with enchantment and ediate between the inner world of consciousness and outer world of ecology and community.[3]

Similar in its message is a collection entitled *Storytelling for Nature Connection Environment, Community and Story-Based Learning* edited by Alida Gersie, Anthony Nanson and Edward Schieffelin with Charlene Collison and Jon Cree (2022). The book 'explores the links between storytelling and emotional literacy, place, environmental justice, connecting with alienated youngsters, how to encourage children and adults' curiosity about nature, building community, sustainability and indigenous peoples, local legends, human-animal communication and how to co-create a sustainable future together'.[4]

High on the rise, there are applicative data-based studies within the ecolinguistic approach. Such is *Corpus-Assisted Ecolinguistics* by Robert Poole (2022):

> Breaking new ground, the book analyzes under-explored environmental discourses that have a tangible impact on ecological wellbeing and sustainability by perpetuating harmful attitudes, practices and ideologies. Chapters present in-depth case studies, including an analysis of the evolving representations of wilderness, an eco-stylistic analysis of a popular novel, and an investigation of the use of humor in reports on animal escapes from slaughterhouses.[5]

Similarly, in *Ecological Stylistics: Ecostylistics Approaches to Discourses of Nature, the Environment and Sustainability*, Daniela Francesca Virdis (2022) explores the instances of the five marker words: 'nature,' 'environment', 'ecosystem', 'ecology' and 'sustainability' on the websites of five big environmental organizations and agencies. Virdis follows Stibbe's (2021) ecosophy, and identifies three key stylistic devices in her analyses: foregrounding, point of view and metaphor (Virdis 2022: 49).

Last but not least, there is the forthcoming publication *Ecolinguistics and Environment in Education: Language, Culture and Textual Analysis* by Emile Bellewes (2024). As the title suggests, the book explores the ecolinguistic potential in teaching environmental issues (cf. Mliless and Larouz 2018).

And this is where we have come with the present volume: cognitive ecostylistics, with its emphasis on the study of texts embedded in socio-economic, environmental and political discourse, looks out into the ecological properties of human thinking and acting as these are indispensable to the story of *Homo cogitans,* i.e. an individual aware of its bodily mental limitations and humbly seeking to establish the harmony with the environment, whether proximal or distal.

The chapters contained in this volume engage the reader in the problem of the present-day Anthropocene manifested as EGO–ECO tensions at the level of communicating self-needs and the needs of the Other. Part One offers an EGO-centric perspective on the relation between the individual, group response, and social system (Chapters 1–5). In Part Two, the viewing frame is zoomed out to embrace the ECO-centric meanings conveyed in texts with the observable shift of focus from the anthropocentric to the anthropoperipheral (Chapters 6–10). In other words, while Part One discusses a human-angled semiotic interplay contained beyond the individual but somewhat intrinsic to the universe of self-centred human engagement, Part Two provides a birds' eye view on the problem of semiotic engagement of texts as extraneous to the human, thereby highlighting the aspects of nature, culture and beyond. In the Introduction to this collection, Stockwell ingeniously sums up this phenomenon of moving out of human self-centredness as 'ripples outwards'.

This EGO–ECO tension is not seen as destructive, however. It rather implies HARMONY, which, itself, is further correlated on an image-schematic level with BALANCE. Tomasz P. Krzeszowski, in his insightful work on the evaluative aspects underlying human conceptualization processes (1997), explores this fundamental BALANCE at length. He argues that 'keeping balance of the body while it acts constitutes such a fundamental experience that we are normally unaware of balance until we lose it and strive to restore it' (Krzeszowski 1997: 126). Of course, we observe this power of BALANCE not only at the level of the individual human experience of action but also at the level of social organization:

> However, social organisation needs to be based on a legal framework if it is not to lapse into anarchy, hence the need for BALANCE that is guaranteed by the vertical line to which traditional leadership and responsibility pertain. Tere is

no other alternative: If choice marks the point at the intersection of the vertical and horizontal dimensions of CUBE, it follows that the hierarchical (individual) and participatory (collective) modes of operation in any organisation are the structural corollaries of the spatio-temporal organisation of reality as a human being experiences it.

(Kuźniak 2021: 188)

BALANCE emerges as a necessary outcome of human operation in the spatiotemporal reality. This means that BALANCE is a conceptual complex composed of such schemas as UP–DOWN, FRONT–BACK, LEFT–RIGHT (cf. Johnson 1987). As Kuźniak (2021) argues, the frame for this conceptualization is CUBE with its inherent IN–OUT polarity characteristic of any CONTAINER. Hence the natural human need to go beyond the limits of their IN (i.e. EGO-centricity) to reach OUT for the ECO (i.e. ECO-centricity). All this resembles a casket: our human bodies are all CONTAINERS operating in the world being the CONTAINER itself, or CUBE, to be exact. The key to resolving the EGO (IN)–ECO (OUT) tension is human awareness of the problem, and this is where cognitive linguistics comes on stage with its philosophy of experientialism and theoretical-methodological tools to systematically explore the secrecies of the human mind.

Elżbieta Mrozek (2015) discusses the above-drawn EGO–ECO tension in terms of the TRANSFER metaphor at a socio-economic plane. She reports, among others, on the findings of the Center for Collective Intelligence Massachusetts Institute of Technology (CCI MIT) about the impact of the development of new digital technologies on the re-shaping of socio-cultural organization and the observed move from the economic pyramid-shaped management towards collective governance. Mrozek (2015: 62–3) quotes McGonigal (2008: 214), who sees collective intelligence as a pivotal factor in the creation of the new forms of knowledge culture grounded in participation rather than individualism. The assumption behind this new thinking is that productivity and efficiency are much higher when they are the product of the synergy of group work rather than dispersed among individuals (cf. Brown 2009).

The turn at EGO–ECO synergy is also manifested at the level of architectural design. Justyna Kleszcz, in her 2020 article 'Challenges of non-anthropocentric architecture: where is the boundary of universal design', explores the issue of bringing engineering solutions regarding the topology of human habitats closer to other species. As she remarks, this type of architecture 'is an approach to perceiving the world, the space surrounding people, which deviates from the paradigm of human superiority in the urban environment, which does not put humans at the absolute centre of all creative activities. It is a way of perceiving space that increases the target group for which the space is designed' (Kleszcz 2020: 33).

On a communicative plane, this reasoning pushes us towards cooperation rather than competition as an optimal strategy in inter-human relations. The milestone in linguistic studies has been laid down in this regard. This is, of course, best illustrated by Paul Grice's (1975) cooperative principle, which asserts that people should rather cooperate than compete in achieving the maximum informative value provided interlocutors obey the so-called maxims of conversation. But the idea behind is ingenious and far-reaching: it is through the human participation, as grounded in day-to-day communication, we are able to gain the maximum of profits. Such a stance does not invalidate individual competitive behaviour as such but sees it as one of the motivational aspects of EGO that otherwise essentially needs ECO to literally survive as a species (cf. Johnson 2010).

The turn towards the appreciation of ECO in linguistics is not only marked by the pragmatics of the 1970s. Parallel to the development of the theory of communication and discourse developed within the varied and not fully compatible paradigms of research comes the birth of cognitive linguistics with its postulate of experientialism and the primacy of the bottom-up, or usage-based, interdisciplinary approach to studies. Naturally, this is the period in which ecological aspects of language study started to gain appreciation among academic circles. It is sufficient to mention here Einar Haugen (1972) and his *The Ecology of Language*, which laid the foundations for the investigation of language in relation to its environment, may safely be labelled a 'hallmark of his work' (see Eliasson 1997; Fill and Mühlhäusler 2001).

It is not accidental that the subject matter of the ecology of language, however then narrowly defined, appears as an important element of the scholarly debate parallel to the emergence of cognitive linguistics (CL). Both approaches share a fundamental imperative to transgress the limitations of the dominant IN-oriented paradigms of language study, i.e. structuralism and generativism, to look for the OUT, contextual aspects. Openness to context has yielded a remarkable potential for the exploration of language seen as embedded in a complicated, dynamic physical–social environment. On a more general plane, it has led to the re-definition of EGO and its participation in the world of communication, not only in relation to other human individuals but other entities that make up its ecology.

Anthropocentricity, in its chauvinistic version, has given in to its perspectival dimension.[6] In this sense, CL is not anti- or non-anthropocentric. Its approach is anthropocentric at heart, but this is perspectivized from the OUTSIDE: the INS cannot exist without their OUTS. The most spectacular effect of that new framing of the IN–OUT relation is definitely seen in the CL departure from the disembodied Cartesian dualism to the so-called embodied mind thesis. As Vyvyan Evans and Melanie Green (2006: 27) note, the thesis holds that 'the human mind and conceptual organization are functions of the ways in which our species-specific bodies interact with the environment we inhabit'. This is the essence of the philosophy of experientialism:

> The idea that experience is embodied entails that we have a species-specific view of the world due to the unique nature of our physical bodies. In other words, our construal of reality is likely to be mediated in large measure by the nature of our bodies ... the nature of our visual apparatus – one aspect of our physical embodiment – determines the nature and range of our visual experience. Similarly, the nature of our biological morphology (the kinds of body parts we have), together with the nature of the physical environment with which we interact, determines other aspects of our experience.
>
> (Evans and Green 2006: 45)

CL has definitely opened up the perspective for the complex investigation of the richness of life hidden behind ECO. EGO, as said above, is still saved but re-defined. So far, CL has marked a clear pathway towards resolving the EGO–ECO tension: the stability of form-meaning relations as expounded in Sausurrian structuralism and Chomskyan generativism has been challenged in favour of the fluidity of meaning with its prototypicality effects (cf. Lakoff and Johnson 1980; Fauconnier 1985; Johnson 1987; Lakoff 1987). The consequence is the move towards richness and complexity. This is the challenge that is contemporarily best illustrated by the research within Fauconnier and Turner's (2002) theory of conceptual integration and Zoltán Kövecses's (2020) extended conceptual metaphor theory.

The challenge for the prospective research is to marry the CL's perspectival anthropocentric ECO-logy with the non-antropocentric philosophy and its posthuman ideological programme, the focus on ontological transformation of the human EGO, or the inclusion of the findings of quantum physics into systematic studies on language, or largely, human communication processes. As it stands, CL, or more precisely, cognitive ecostylistics, which is one of the most dynamically flourishing offspring of the CL approach, and the focus of which is the present volume, has hopefully delineated its area of exploration – the pathway from EGOcentrism to ECOcentrism at the level of language awareness (Chapters 1–5), social issues (Chapters 6–9) and environmental concerns (Chapters 1, 5 and 10).

Notes

1. Translated with DeepL.com (the original passage in Polish).
2. 'The term has no single definition. Rather, it has come to represent a vibrant and unruly spectrum of transdisciplinary approaches that are unified by a common and deceptively simple argument: attending to worlds that are more than human requires changing the methods and apparatuses of study. In other words, to dissolve the boundary between nature and culture is to radically remix the arts, humanities, and the social and natural sciences' (https://as.nyu.edu/departments/xe/curriculum/past-semester-courses/courses-spring-2019/Nomenclature.html [accessed 24 January 2023]).
3. https://www.bloomsbury.com/uk/storytelling-and-ecology-9781350114920 (accessed 25 January 2023).
4. https://www.hawthornpress.com/books/storytelling/storytelling-for-nature-connection/ (accessed 25 January 2023).
5. https://www.bloomsbury.com/uk/corpusassisted-ecolinguistics-9781350138551/ (accessed 26 January 2023).
6. See Drewniok and Kuźniak's Chapter 1 in the present volume for a more detailed discussion on the aspects of anthropocentricity in language and philosophy.

References

Aloi, G. and S. McHugh, eds (2021a), *Posthumanism in Art and Science*, New York: Columbia University Press.

Aloi, G. and S. McHugh (2021b), 'Envisioning posthumanism', in G. Aloi and S. McHugh (eds), *Posthumanism in Art and Science*, 1–21. New York: Columbia University Press.

Bellewes, E. (2024, forthcoming). *Ecolinguistics and Environment in Education: Language, Culture and Textual Analysis*. London: Bloomsbury.

Berger, J. (2005), 'Why look at animals? A close reading', *World Views Environment Culture Religion*, 9 (2): 203–18.

Bogusławska, M. (2022), 'Introduction. Ecolinguistics in the new millennium (noted in the year 2022)', in M. Bogusławska, A. A. Dragoescu Urlica, L. Kamberi (eds), *From Cognitivism to Ecologism in Language Studies*, 9–16. Berlin: Peter Lang.

Bogusławska, M. A. A Dragoescu Urlica and L. Kamberi, eds (2022), *From Cognitivism to Ecologism in Language Studies*, Berlin: Peter Lang.

Brown, T. (2009), *Change by Design: How Design Thinking Transforms Organizations and Inspires Innovation*, New York: HarperCollins Publishers.

Derrida, J. (2002), 'The animal that therefore I am (more to follow)', trans. David Wills. *Critical Inquiry*, 28 (2): 369–418.

Descola, P. (2014), 'Beyond nature and culture', in G. Harvey (ed.), *The Handbook of Contemporary Animism*, 77–91. London: Routledge.

Eliasson, S. (1997), 'Einar Haugen (1906–1994)', in S. Eliasson and E. H. Jahr (eds), *Language and Its Ecology. Essays in Memory of Einar Haugen* [Trends in Linguistics. Studies and Monographs, 100], v–xi. Berlin: Mouton de Gruyter.

Evans, V. and M. Green (2006), *Cognitive Linguistics. An Introduction*, Edinburgh: Edinburgh University Press.

Fauconnier, G. (1985), *Mental Spaces*, Cambridge: Cambridge University Press.

Fauconnier, G. and M. Turner (2002), *The Way We Think: Conceptual Blending and the Mind's Hidden Complexities*, New York: Basic Books.

Foucault, M. (2002/1966), *The Order of Things: An Archaeology of the Human Sciences* [Les mots et les choses: Une archéologie des sciences humaines], London: Routledge.

Gersie, A., A. Nanson, E. Schieffelin et al., eds (2022), *Storytelling for Nature Connection Environment, Community and Story-Based Learning*, Stroud: Hawthorn Press.

Grice, H. P. (1975), 'Logic and conversation', in P. Cole and J. Morgan (eds), *Syntax and Semantics*, 41–58. New York: Academic Press.

Haraway, D. (2013), 'A cyborg manifesto: Science, technology, and socialist-feminism in the late twentieth century', in S. Stryker and S. Whittle (eds), *The Transgender Studies Reader*, 103–18. London: Routledge.

Haugen, E. (1972), *The Ecology of Language*, Stanford: Stanford University Press.

Hoły-Łuczaj, M. (2018), *Radykalny nonantropocentryzm. Martin Heidegger i ekologia głęboka*, Warszawa: Wydawnictwo Uniwersytetu Warszawskiego.

https://www.deepl.com/translator (accessed 25 January 2023).

https://as.nyu.edu/departments/xe/curriculum/past-semester-courses/courses-spring-2019/Nomenclature.html (accessed 24 January 2023).

https://www.bloomsbury.com/uk/storytelling-and-ecology-9781350114920 (accessed 25 January 2023).

https://www.hawthornpress.com/books/storytelling/storytelling-for-nature-connection/ (accessed 25 January 2023).

https://www.bloomsbury.com/uk/corpusassisted-ecolinguistics-9781350138551/ (accessed 26 January 2023).

Fill, A. and P. Mühlhäusler, eds (2001), *Ecolinguistics Reader: Language, Ecology and Environment*, London: Continuum.

Johnson, M. (1987), *The Body in the Mind: The Bodily Basis of Cognition*, Chicago: University of Chicago Press.

Johnson, S. (2010), *Where the Good Ideas Come From. The Natural History of Innovation*, New York: Riverhead Books.

Kleszcz, J. (2020), 'Challenges of non-anthropocentric architecture: Where is the boundary of universal design', *Housing Environment 31/2020*, 33–41. Available online: https://www.ejournals.eu/housingenvironment/2020/31-2020/art/18005/ (accessed 7 February 2023).

Kövecses, Z. (2020), *Extended Conceptual Metaphor Theory*, Cambridge: Cambridge University Press.

Krzeszowski, T. P. (1997), *Angels and Devils in Hell. Elements of Axiology in Semantics*, Warszawa: Wydawnictwo Energeia.

Kuźniak, M. (2021), *The Geometry of Choice: Language, Culture, and Education*, Cham: Palgrave Macmillan.

Lakoff, G. (1987), *Women, Fire, and Dangerous Things. What Categories Reveal about the Mind*, Chicago: University of Chicago Press.

Lakoff, G. and M. Johnson (1980), *Metaphors We Live By*, Chicago: University of Chicago Press.

McGonigal, J. (2008), 'Why I love bees: A case study in collective intelligence gaming', in K. Salen (ed.), *The Ecology of Games: Connecting Youth, Games, and Learning*, Cambridge, Massachusetts: MIT Press.

Mliless, M. and M. Larouz (2018), 'An ecolinguistic analysis of environment texts in Moroccan English language teaching textbooks', *International Journal of Environmental Studies*, 5: 103–16.

Mrozek, E. (2015), 'Kolektywna inteligencja i partycypacyjna kultura podejmowania decyzji', in A. Dąbrowski, A. Schumann and J. Woleński (eds), *Podejmowanie decyzji. Pojęcia, teorie, kontrowersje*, 51–88. Kraków: Copernicus Center Press.

Nanson, A. (2021), *Storytelling and Ecology: Empathy, Enchantment and Emergence in the Use of Oral Narratives*, [Bloomsbury Advances in Ecolinguistics]. London: Bloomsbury.

Poole, R. (2022), *Corpus-Assisted Ecolinguistics* [Bloomsbury Advances in Ecolinguistics], London: Bloomsbury.

Stibbe, A. (2021), *Ecolinguistics: Language, Ecology and the Stories We Live By*, 2nd edn, Abingdon: Routledge.

Virdis, D.F. (2022), *Ecological Stylistics: Ecostylistic Approaches to Discourses of Nature, the Environment and Sustainability*, Cham: Palgrave Macmillan.

Wolfe, C. (2010), *What Is Posthumanism?*, Minnesota: University of Minnesota Press.

Yoo, D-S. (n.d.), 'Organ-machine hybrid: Experiments in combinations of animal organs with electronic devices and robotics for new artistic applications', *Journal of the New Media Caucus*. Available online: http://median.newmediacaucus.org/isea2012-machine-wilderness/organ-machine-hybrid-experiments-in-combinations-of-animal-organs-with-electronic-devices-and-robotics-for-new-artisticapplications/ (accessed 1 February 2023).

Index

Abbyy FineReader application 125–6
action chain 141–4, 147, 182
adaptation-oriented translation strategy 156
aesthetics
    aesthetic experience 14, 101, 103–4, 109–10, 180
    aesthetic information processing 104, 107
    embodied 101–4, 110–11
African American Diaspora 156, 163
African American Vernacular English (AAVE) 156–7
alliteration 6–7, 65
Aloi, G., *Posthumanism in Art and Science* 209
Amazon, reviews 35–7, 41
Andalusian Plan for Research Development and Innovation 128
Anglo-Saxon 154, 167
Anne of Green Gables (Montgomery) 14, 100
    Anne Shirley (fictional character) 104–6, 108
    landscape studies of (*see* landscape studies)
    Matthew Cuthbert (fictional character) 106–7
    Polish translations of 106–11
    Rozalia Bernsteinowa (fictional character) 106, 108–9
antagonism/antagonistic 2, 189
Antconc software 54, 125–7
Anthropocene 209, 211
anthropocentric/anthropocentrism 15, 20, 30 n.2, 209, 214. *See also* non-anthropocentric/-anthropocentrism
    chauvinistic 20, 30 n.2
    perspectival 17, 20, 30 n.2
applied ecostylistics 2, 116
Appraisal Theory 37
Atwood, M. *See Cat's Eye*

Bakhtin, M. M.
    on dialogism 43
    primary *vs.* secondary genre dichotomy 21–2
    on speech genres 20, 22
Bańkowska, A., Polish retranslation of *Anne of Green Gables* 106–11
Barad, K. M. 34–5, 47
Barbusse, H., *Le Feu* 138–9
Barnes, J., *TSOAE. See The Sense of an Ending* (*TSOAE*)
Berger, J., *Why Look at Animals?* 209
Biden, J., analysis of victory speech 54–5
Black Semantics 157
blurbs (short texts) 13, 16, 20–2, 29–30. *See also* hotel blurbs
bodily context 20
bodily movements 116, 191–2, 195–7, 200–1
Bogusławska, M., *From Cognitivism to Ecologism in Language Studies* 17, 208
book reviewing genre 3, 36, 39, 47
Borden, M.
    The Forbidden Zone 138
    'Song of the Mud' 138
boulomaic modal-world 44–5, 47
British National Corpus (BNC) 56
Brown, D. A. 192–3

Camus, A., *The Fall. See The Fall* (*TF*)
canonical event model 141
Caracciolo, M. 191–2
    literary analysis 195
Cat's Eye (Atwood) 115, 171
    Cordelia (fictional character) 171–2
    Elaine (fictional character) 115–16, 171–2, 178, 185
    mind style of 172, 174–7, 179, 182, 185
    perception of self 177–80
    stylistic presentation of still life/self-harm 180–4

epistemic modal-world 182–4
fictional ekphrasis in 173–5
grounding strategies 181
Center for Collective Intelligence Massachusetts Institute of Technology (CCI MIT) 212
chauvinistic anthropocentrism 20, 30 n.2. *See also* perspectival anthropocentrism
class inequality 117–20, 124
cognitive-conceptual context 20, 22, 26
cognitive discourse grammar 174
cognitive experiences of readers 99–100
Cognitive Grammar (CG) 7, 141, 144–5, 149, 172, 174–5, 185
cognitive linguistics (CL) approach 2–3, 5, 13, 15–19, 128, 207, 212–14
cognitive poetics 5, 7, 172–3
cognitive psychology 5
cognitive science 99, 101, 195
cognitive sensorium 5, 105
cognitive stylistics 2, 5, 70, 136, 173, 175
Coleridge, S. T.
   Notebook 24 6–8
   The Rime of the Ancient Mariner 8–9
compléments cognitifs tool 159, 167
conceptualization 16, 19–20, 22, 173, 175, 179–81, 184, 190, 211–12
conceptual metaphor theory (CMT) 3, 14, 18–19, 30, 44, 46–7, 69–70, 75, 89, 116, 176, 191, 198–9, 214
   cross-domain mappings 198–9
   domain level 19, 21, 25
   frame level 19, 21–2
   image-schema level 19
   mental-space level 19–22, 26, 28
Condé Nast Traveller magazine 13, 16, 20–2, 29
   boast cultural inspirations 24
   hotel blurbs 24, 26–7
   media kit 27–8
   target recipient of 27–9
   website 27–8
construal 7, 19, 47, 142, 145, 174–5, 178–9, 181
content analysis 53–4
contrastive analysis 153, 165
conventional stylistic analysis 2–3, 9, 175
corpus-assisted discourse studies (CADS) 118–22, 126

corpus linguistics (CL) 4, 54, 120
The Corpus of News on Economic Inequality (1971-2020) 119, 122
   compilation process 124–5
   corpus inspection 125–7
   data 122–4
corpus stylistics 4–5
Covid-19 pandemic 13, 16, 27, 30, 51–2, 64
   health communication (*see* health communication)
   optimistic anxiety in 52, 61
   structural form of speeches on 59–60
   use of poetry in speech 63–5
Crane, M. T. 100
Critical Discourse Analysis (CDA) 2, 53, 55, 119–21, 128
Critical Discourse Studies journal 119
Crusie, J., *Bet me* 191
cubism 209
cultural-language symbols 159, 167

Das, S. 135, 137–8
The Daylight Gate (TDG), (Winterson) 37, 39–40, 48
deep ecology 17–19, 30 nn.2–3, 30 n.5. *See also* shallow ecology
   principles of 208
deictic 9, 29, 38, 41, 44, 116, 182, 191, 195, 197–8, 201
   deictic projection 197, 201
Derrida, J., *The Animal That Therefore I am* 209
Descola, P. 209
Devall, B. 31 n. 5
diachronic 8, 16, 118, 120, 127
Dialogic Syntax (DS) 37, 43
diffractive approach 13, 33–7, 47–8
   diffraction in ocean waves 36
   propagation of wave through chink 35
DINEQ (*Discourse* and *INEQuality*) corpus 122–3, 125–7
   compilation process of 122–4
Dionysus of Halicarnassus 190
direct speech 41, 43
discourse context 20, 26
Discourse & Society journal 119
discourse-world 37–41, 44–5, 47
dOCUMENTA art exhibition (2012) 209

domestication (*Einbürgerung*) 154, 156, 158, 165
Dragoescu Urlica, A. A., *From Cognitivism to Ecologism in Language Studies* 17, 208

eco-centricity 16, 212
ecology 2, 6, 8–9, 115, 207, 211, 213–14
  deep ecology 17–19, 30 n.2–3, 30 n.5
  ecological crisis 17, 30 n. 3, 210
  ecological stylistics 3, 5–9, 207
  shallow ecology 17–18, 30 n.3
economic inequality 115, 117–18, 122, 124, 127–8
  in British 117–20
eco-philosophy/eco-philosopher 16–17
ecosophy 9, 210
eco-sphere 18, 20
ecostylistics 2–3, 8–9, 135, 149, 165, 173, 209–11
ecosystem 13, 27, 30, 115, 210
eco-translation 115, 156, 158–9, 165, 168
ego-centric/egocentricity 1, 4–5, 15, 17, 33, 207, 209, 211–12
ego-eco continuum 15–20, 24–5, 27, 29–30, 33–4, 208–9, 211–15
ekphrasis 115, 172, 177, 185, 189–90, 202
  fictional ekphrasis 173–5
embedded cognition 5–6
embodied aesthetics 101–4, 110–11
embodied cognition 5–6, 177, 191
embodied simulation 101–3, 192, 195
embodiment paradigm 101, 103, 192
empathetic engagement 69–71, 74, 84, 86–7, 91, 91 n.2
empathetic recognizability 200
empathy 13–14, 52, 57, 64, 69, 85–6, 89, 91 n.2, 91 n.4
enactive cognition 5–6, 191
enargeia (vividness) 190, 192, 202
endo-linguistics 18
The Equality Trust (2022) 124, 128
ERC grants 128
Evans, V., philosophy of experientialism 214
exo-linguistics 18
exophoric reference 38, 43
experimental data collection approach 73–4. *See also* naturalistic data collection approach

extended cognition 5–6
Extended Conceptual Metaphor Theory (ECMT) 13, 19–20, 207
extralinguistic communication 166

Fairclough, N. 55, 63
The Fall (*TF*), (Camus) 37–40, 48
Farmer, E., *Ecolinguistics and Environment in Education: Language, Culture and Textual Analysis* 211
feminist stylistics 2–3
Fernandez-Quintanilla, C. 69, 71–2, 84, 86, 89–91
  implicit-evidence-for-empathy codes 96–7
figurative language 13, 53–5, 65–6
  in ICSC 58–62
  Leech on categories of 55, 60
  public response to the use of 62–4
The Flowers (Walker) 69, 73–4, 91
  analysing pathetic fallacy in 75–83
  Myop (fictional character) 75–7, 83–9
  participants' think-aloud responses 78–88
  scoring system 77, 83
'For a Venetian Pastoral by Giorgione (In the Louvre)' (Rossetti) 116, 192
  bodily movements 116, 191–2, 195–7, 201
  CMT 198–200
  deixis and feeling of presence 197–8
  enactment and bodily engagement 195–7
  image schema 200–202
  negative accommodation 196
  pastoral tradition and Rossetti's sonnet 192–5
  perceptual deixis 197–8
  personification 195, 199–200
  temporal deixis 198
foreignization (*Verfremdung*) 156
formalists 3–5
Foucault, M., *The Order of Things* 209
4E cognition 5–8, 207. *See also specific cognitions*
frame system theory 103
'Futility', landscape of (Owen) 115, 135, 139–45, 149–50
  human body and landscape 135–9
  interpretations of 145–9

landscapes of First World War 136–9
single-participant process 141
thematic process 141, 143–4

Gallese, V. 102–3
generativism 213–14. *See also*
  structuralism
geographical metaphors 100
geopolitical/geographical transformation
  99–100
The Germ magazine 192
Gersie, A., *Storytelling for Nature
  Connection Environment,
  Community and Story-Based
  Learning* 210
Goethe 33
Goodreads 73, 191
Great Chain of Being 18
Green, M., philosophy of experientialism
  214

Haraway, D., *A Cyborg Manifesto* 209
Haraway, D. J. 33
Haugen, E., *The Ecology of Language* 213
health communication, Covid-19 51–2, 57
  guidance/guidelines 52, 65
Heidegger, M., *Dasein* 33
Helsinger, E. 194–5
hierarchical ontology of beings 18–19,
  30 n.4
Hip Hop language/street speech 157
Hoły-Łuczaj, M. 16–17, 30 n. 5, 208
homelessness 120
homocentrist 16
homo cogitans 211
homodiegetic narrator 38, 41–4
hotel blurbs 16, 21–5, 29–30
  description of Cornish Ukiyo 24
  genres 21–3
  grammar 26–7
  lexis 24–6
  target recipient of 27–9
  themes 23–4
Hothouse Flowers (*HF*), (Riley) 42, 45–6, 48
  Harry (fictional character) 45
  Julia (fictional character) 46
  Olivia (fictional character) 45
human conceptualizer/conceptualization
  15, 17, 211
hyperbole/hyperbolic expression 55, 60

idealist 16
idiosyncratic 9, 48, 102, 146, 159, 172–3,
  175, 184–5
income inequality 117, 124
individual experience 27, 72, 102–3, 207
intercultural communication 153
interdiscipline/interdisciplinary 99, 101,
  111, 213
interdiscursivity 58–9
interlingual translation act 153–4
interpersonal communication 159–60,
  166
intertextuality/intertextual references
  40–1, 62, 66, 77, 84
intra-actions 34–48
intralingual translation act 115, 153–6,
  158–64, 167
intratextual 174, 181–2, 184–5
Ireland/Irish 51, 202
  ceremonial language 62
  ICSC (*see* Irish Covid Speeches Corpus
    (ICSC))
  Irish context of Covid-19 public health
    discourse 52–3
  linguistic characterization of corpus
    53–4
  National Public Health Emergency
    Team 56
  Official Languages Act 2003 62
  structural form of speeches on
    Covid-19 59–60
  Varadkar's speech on lockdown 51, 53
Irish Covid Speeches Corpus (ICSC)
  53–5
  concordance of *one of us* in 58
  figurative language in 58–62, 65
    occurrences of figurative
      expressions in ICSC 60
    public response to the use of
      62–4
  frequencies of pronouns used in
    57–8
  salient characteristics of 56–8
  top keywords in 56
  top three-word units in 56–7
Ishiguro, K. *Never Let me Go. See Never
  Let me Go* (NLMG)

Johnson, M., *Metaphors We Live By* 198
Journal of Literary Semantics 91 n. 3

Kamberi, L., *From Cognitivism to Ecologism in Language Studies* 17, 208
Kleszcz, J., 'Challenges of non-anthropocentric architecture: where is the boundary of universal design' 212
Kövecses, Z. 13, 19, 214
  on schematicity hierarchy 19–20
  on types of context 20, 24
Krzeszowski, T. P. 18, 211
Kuijpers, M. M. 191
Kuzmičová, A. 191, 195–6, 201

Lakoff, G., *Metaphors We Live By* 198
landscape studies
  of *Anne of Green Gables* 14, 99–101
    aesthetics embodiment 101
    methodology 103–10
    'Futility' (*see* 'Futility', landscape of)
Langacker, R. 141
language comprehension 101, 191
Le Concert Champêtre (Concert in the Open Air) painting 192, 202
Leech, G. N., on categories of figurative language 55, 60, 175
  hyperbole 55, 60
  metaphor 55, 60–1, 64
  personification 55, 60–1
  simile 55, 60
Lee, T., *Macbeth*. *See Macbeth*
lexical nodes 104, 111–12 n.1
lexico-grammatical tools 4
lexicographical analysis 157
linguistic description of text 1–2, 4
linguistic expressions 197–8
linguistic representation 135, 149, 172, 175
literal language 55
literary critic/criticism 1, 3–4, 38, 148, 150 n.1
Literary Response Questionnaire 74
literary texts 2, 4, 6, 33, 100, 115, 154, 156, 159, 165, 172–3, 185, 189, 191–2, 195, 197
  and response (reader-response) 33–5, 38, 48 (*see also specific novels*)
literary translation 35, 99–100, 110–11, 156, 159–60
Lord, M., poetry pandemic 63

Lovejoy, A. O., *The Great Chain of Being: A Study of the History of an Idea* 18
luxury travel 27–30

Macbeth (Lee) 115, 153–9
  adaptation technique of compensation 162
  adaptation technique of enrichment 163
  adaptation technique of equivalence 161
  adjective distribution 166–7
  log-likelihood keyness corpus analysis of 164
  slang, urban (African American) 158–9, 166
    exploitation of translation procedures 159–63
    use of contractions 165–6
    use of gerunds/continuous verbs/conjuncts 164–5
Marković, S. 104
Martin, M. 51, 53
Mason, J., theory of narrative interrelation 174
Massey, D., on space 8
McGhee, F. 194–6
McHugh, S., *Posthumanism in Art and Science* 209
mental representation 14, 43, 83–4, 88, 90
metaphor/metaphorical expression 55, 60–1, 64, 198. *See also* conceptual metaphor theory (CMT)
metastasis 190
metathesis 190
Miall, D. 191
mind-modelling approach 7–9, 148, 180
Minsky, M. 103, 112 n.1
monoglossia 42
Montgomery, L. M. *See Anne of Green Gables*
Mrozek, E. 212
multifaceted infotainment chunks 22
multiperspectivity 101

Naess, A. 9, 16, 19, 208
  ecological crisis 17, 30 n.3
  'The Shallow and the Deep, Long-Range Ecology Movement' 17, 30 n.3

Nanson, Anthony, *Storytelling and Ecology: Empathy, Enchantment and Emergence in the Use of Oral Narratives* 208
narrative empathy 13, 69–70, 91 n.3, 91 n.5
   analysis of 83–5
   analytical methodologies for PF perception and 74–7, 83
   data collection circumstances 73–4
      experimental 73–4
      naturalistic 73–4
   data requirements 72–3
   data summary 85–6, 88–9
   empathetic responses on reading story 72
   impact of narrative perspective on (study) 71
   stylistic exploration of 70–1, 91 n.3
   and think-aloud protocols 71–4
naturalistic data collection approach 73–4. *See also* experimental data collection approach
neoliberalism 118
neuroscience 101, 103
Never Let me Go (NLMG), (Ishiguro) 42–4, 46, 48
   Katy H. Rather (fictional character) 43
Nexis Uni online database 123–5, 127–8
   access to UK national newspapers through 124
Nikolajeva, M. 104
non-anthropocentric/-anthropocentrism 15–16, 208–9, 212, 214. *See also* anthropocentric/anthropocentrism
non-empathy 85, 88
non-sensorium analysis 105. *See also* sensorium analysis
non-verbal data 72–3. *See also* verbal data

O'Brien, E., 'The Anxiety of Influence' 65
observer's paradox 34
offline processing data 76, 91 n.1. *See also* online processing data
Olejniczak, J. 168 n.1
omniscient narrator 45
online book reviewing genre 13, 34–6, 191
online processing data 76, 90, 91 n.1. *See also* offline processing data

online socio-material environment 36, 39
ontology 16, 18, 33, 207–9
   ontological hierarchy of beings (*see* hierarchical ontology of beings)
optical character recognition (OCR) technology 123, 125
origo 116, 177, 197
Owen, W.
   'Dulce et Decorum Est' 137–8
   'Futility' (*see* 'Futility', landscape of)

Palmer, A., social minds 6
pathetic fallacy (PF) 13, 69–70, 73
   analysing PF in *The Flowers* (Walker) 74–83
   effects of PF in narratives 75
   linguistic indicators of 75
   mappings 75–6, 83, 85–6, 88–9
pay inequality 117, 124
perceptual experience 103, 111 n.1, 192, 195
personification expression 55, 60–1, 116, 195, 199–200, 202
perspectival anthropocentrism 17, 20, 30 n.2. *See also* chauvinistic anthropocentrism
poetic persona 7, 9
Poetics and Linguistics Association (PALA) 21
poetry pandemic 63
Poole, R., *Corpus-Assisted Ecolinguistics* 210
poverty 109, 119
pragmatics 4, 104, 107, 109–10, 213
presumed intimacy 39–40, 47

quantum physics 208–9, 214
Quintilian 190

Raab, D. 64
radical ecological analysis 6, 9, 207
Rajandran, K., analysis of metaphors for Covid-19 61
Raping/MCing 156
reader-response stylistics 1, 5, 13, 20, 33–5, 38, 48, 70–4, 89–90, 136, 146
   narrative empathy (*see* narrative empathy)
   one-to-many relationships in 34
   online reader response data 146

reference corpus 4–5, 54, 56
Renaissance 192–4
research questions (RQs) 69–70, 73, 121, 154, 167
retranslations 100–1, 106–11
rhetoric/rhetorical 2–3, 36, 38, 65, 144, 154, 166, 169–70, 173, 179, 190, 207
Riley, L., *Hothouse Flowers*. *See* Hothouse Flowers (HF)
Rojek, C., on presumed intimacy 39–40, 47
Rosch, E., *The Embodied Mind: Cognitive Science and Human Experience* 195
Rossetti, D. G., 'For a Venetian Pastoral by Giorgione'. *See* 'For a Venetian Pastoral by Giorgione (In the Louvre)'

Sannazaro, J., *Arcadia* 193
Saramago, J., *Blindness* 195
schematicity hierarchy 19
Schleiermacher, F. 156, 158, 165
self-centred stylistics 1–3, 211
The Sense of an Ending (*TSOAE*) (Barnes), online reviews of 13, 34, 36–9, 42–3, 48
   diffractive approach 34–7
   focal resonance 43
   reader's (R1 and R2) review of 36
      difference and pass-wor(l)ds 42–7
      intertextuality in 40–1
      repetition and order-wor(l)ds 37–42
   Tony Webster (narrator) 38, 41–3, 46–7
   Veronica Ford (fictional character) 41–5, 47
sensorimotor 101, 103, 111 n.1, 116, 191, 195
sensorium analysis 104–5. *See also* non-sensorium analysis
Shakespear, W., *Macbeth*. *See* Macbeth
shallow ecology 17–18, 30 n.3. *See also* deep ecology
short texts. *See* blurbs (short texts)
simile expression 55, 60, 138, 184
simulationism 102, 105
situational context 20, 30
Sketch Engine tool 125

slang, urban (African American) 158–9, 166
   exploitation of translation procedures of *Macbeth* 159–63
social class 118
social constructionism 118
social inequality 124
social media 55, 63, 119, 122, 209
social variables 8, 119, 127
socio-cultural conditions 2, 15, 99, 121, 155–6, 159, 167, 173, 212
sociolinguistics 4–5, 156–7, 163, 168
   slang in 158–9
The Spanish Research Agency National Plan 128
specificity 103, 110, 115, 142, 156, 175, 181, 183–4
speech genres 20, 22
Stevens, M. 28
storytelling 29, 165, 177–8, 210
street speech, African American 154, 156–7, 165, 167–8
structuralism 213–14. *See also* generativism
stylistic analysis 2–6, 9, 37, 46, 135
survey questions 94–5
   matrix table for 96
sustainability 2, 27–9, 210
systemic-functional approach 2, 4

target language 153–4, 156, 165
target recipients 27–9, 107
Text World Theory (TWT) 37–8, 40–1, 43–5, 172, 174, 176–7, 185, 191
Thani, D. 28
Theocritus, *Idylls* 193
think-aloud protocol 69–70
   affordances of think-aloud data 89–90
   and narrative empathy 71–4
   participants' responses 79–82
Thompson, E., *The Embodied Mind: Cognitive Science and Human Experience* 195
Titian 192–3
Toolan, M. 120, 122
traditional stylistics 1–2, 6, 207
transcendence 8, 30 n.2, 44, 104, 116, 177, 202
transitivity 2, 121, 127, 195–6
translational stylistics approach 14, 101, 110–11

translation studies 99–101, 109–11, 153.
    See also Anne of Green Gables
translator's style 101
transportational potential of literary text 192, 201–2
trigger-lexical items 111 n.1
Twilight novel 191
Twitter 55, 63–4

UAM Corpus Tool 125
The United Kingdom 118
    Audit Bureau of Circulations 123
    economic inequality 117–20
The United States 118, 163

vaghezza (vagueness) 194
Varadkar, L.
    speech on lockdown 51, 53, 55
    speech reference on Seamus Heaney 55, 62–4
Varela, F. J., *The Embodied Mind: Cognitive Science and Human Experience* 195

verbal data 72–3. *See also* non-verbal data
verbal representation 187
Virdis, D. F., *Ecological Stylistics: Ecostylistic Approaches to Discourses of Nature, the Environment and Sustainability* 3, 210
Virgil, *Eclogues* 193
visual representation 189

Walker, A., *The Flowers* 69, 73–4
wealth inequality 117, 120, 124
West, A. G., 'God! How I Hate You, You Cheerful Young Men' 136–7
White Sky 22
'Wilfred Owen Futility analysis' 146
Winterson, J., *The Daylight Gate*. See *The Daylight Gate* (*TDG*)
Wolfe, C., *What Is Posthumanism* 209
World Inequality Report (2022) 117
world-switch 44, 174

Yeats, W. B. 65
Yoo, D-S., *Organ-Machine Hybrid* 209

www.ingramcontent.com/pod-product-compliance
Lightning Source LLC
Chambersburg PA
CBHW071832300426
44116CB00009B/1524